Course
Part B **4**

Contemporary Mathematics in Context
A Unified Approach

CORE-PLUS MATHEMATICS PROJECT

Course 4
Part B

Contemporary Mathematics in Context
A Unified Approach

Arthur F. Coxford
James T. Fey
Christian R. Hirsch
Harold L. Schoen
Eric W. Hart
Brian A. Keller
Ann E. Watkins
with
Beth E. Ritsema
Rebecca K. Walker

Glencoe McGraw-Hill

New York, New York   Columbus, Ohio   Chicago, Illinois   Peoria, Illinois   Woodland Hills, California

Glencoe/McGraw-Hill

*A Division of The **McGraw·Hill** Companies*

This project was supported, in part, by the National Science Foundation.
The opinions expressed are those of the authors and not necessarily those of the Foundation.

Send all inquiries to:
Glencoe/McGraw-Hill
8787 Orion Place
Columbus, OH 43240-4027

ISBN 0-07-827549-0 (Part A)
ISBN 0-07-827550-4 (Part B)

Contemporary Mathematics in Context
Course 4 Part B Student Edition

2   3   4   5   6   7   8   9   10   004/004   10 09 08 07 06 05 04 03

Dedication

The authors, teachers, and publication staff who have collaborated in developing *Contemporary Mathematics in Context* dedicate this fourth course in the series to the memory of our friend and professional colleague Arthur Coxford, who passed away on March 5, 2000. Professor Coxford's breadth of knowledge and experience in mathematics education, his consistent good judgment and hard work, and his good-humored leadership were instrumental in all aspects of the Core-Plus Mathematics Project. We will miss his thoughtful contributions to project activities, but he has left an invaluable legacy of creative work and a striking example of personal and professional commitment and integrity.

Core-Plus Mathematics Project Development Team

Project Directors
Christian R. Hirsch
Western Michigan University
Arthur F. Coxford
University of Michigan
James T. Fey
University of Maryland
Harold L. Schoen
University of Iowa

Senior Curriculum Developers
Eric W. Hart
Maharishi University of Management
Brian A. Keller
Michigan State University
Ann E. Watkins
California State University, Northridge

Professional Development Coordinator
Beth E. Ritsema
Western Michigan University

Evaluation Coordinator
Steven W. Ziebarth
Western Michigan University

Advisory Board
Diane Briars
Pittsburgh Public Schools
Gail Burrill
University of Wisconsin–Madison

Jeremy Kilpatrick
University of Georgia
Kenneth Ruthven
University of Cambridge
David A. Smith
Duke University
Edna Vasquez
Detroit Renaissance High School

Curriculum Development Consultants
Kenneth A. Ross
University of Oregon
Richard Scheaffer
University of Florida
Paul Zorn
St. Olaf College

Technical Coordinator
James Laser
Western Michigan University

Collaborating Teachers
Emma Ames
Oakland Mills High School, Maryland
Mary Jo Messenger
Howard County Public Schools, Maryland
Valerie Mills
Ann Arbor Public Schools, Michigan

Graduate Assistants
Cos Fi
University of Iowa

Sarah Field
University of Iowa
Kelly Finn
University of Iowa
Gina Garza-Kling
Western Michigan University
Chris Rasmussen
University of Maryland
Heather Thompson
Iowa State University
Roberto Villarubi
University of Maryland
Rebecca Walker
Western Michigan University
Edward Wall
University of Michigan
Marcia Weinhold
Western Michigan University

Production and Support Staff
Kelly MacLean
Angela Reiter
Anna Seif
Wendy Weaver
Kathryn Wright
Teresa Ziebarth
Western Michigan University
Catherine Kern
University of Iowa

Core-Plus Mathematics Project Field-Test Sites

Special thanks are extended to these teachers and their students who participated in the testing and evaluation of Course 4.

Ann Arbor Huron High School
Ann Arbor, Michigan
 Ginger Gajar
 Brenda Garr

Ann Arbor Pioneer High School
Ann Arbor, Michigan
 Jim Brink

Arthur Hill High School
Saginaw, Michigan
 Virginia Abbott
 Cindy Bosco

Battle Creek Central High School
Battle Creek, Michigan
 Teresa Ballard
 Steven Ohs

Battle Creek Mathematics & Science Center
Battle Creek, Michigan
 Dana Johnson
 Serena Kershner
 Rose Martin
 Lily Nordmoe

Bedford High School
Temperance, Michigan
 Ellen Bacon
 David J. DeGrace

Bloomfield Hills Andover High School
Bloomfield Hills, Michigan
 Jane Briskey
 Cathy King
 Linda Robinson
 Mike Shelly

Brookwood High School
Snellville, Georgia
 Ginny Hanley

Caledonia High School
Caledonia, Michigan
 Jenny Diekevers
 Gerard Wagner

Centaurus High School
Lafayette, Colorado
 Dana Hodel
 Gail Reichert

Clio High School
Clio, Michigan
 Bruce Hanson
 Lee Sheridan
 Paul Webster

Davison High School
Davison, Michigan
 John Bale
 Tammy Heavner

Ellet High School
Akron, Ohio
 Marcia Csipke
 Jim Fillmore

Firestone High School
Akron, Ohio
 Barbara Crucs

Goodrich High School
Goodrich, Michigan
 John Doerr
 Barbara Ravas
 Bonnie Stojek

Grand Blanc High School
Grand Blanc, Michigan
  Charles Carmody
  Linda Nielsen

Grass Lake Junior/Senior High School
Grass Lake, Michigan
  Brad Coffey

Kelloggsville Public Schools
Wyoming, Michigan
  Steve Ramsey

Lakeview High School
Battle Creek, Michigan
  Larry Laughlin
  Bob O'Connor
  Donna Wells

Midland Valley High School
Langley, South Carolina
  Ron Bell
  Janice Lee

North Lamar High School
Paris, Texas
  Tommy Eads
  Barbara Eatherly

Okemos High School
Okemos, Michigan
  Lisa Magee
  Jacqueline Stewart

Portage Northern High School
Portage, Michigan
  Renee Esper
  Pete Jarrad
  Scott Moore

Prairie High School
Cedar Rapids, Iowa
  Judy Slezak

San Pasqual High School
Escondido, California
  Damon Blackman
  Ron Peet

Sitka High School
Sitka, Alaska
  Cheryl Bach Hedden
  Dan Langbauer

Sturgis High School
Sturgis, Michigan
  Craig Evans
  Kathy Parkhurst

Sweetwater High School
National City, California
  Bill Bokesch

Tecumseh High School
Tecumseh, Michigan
  Jennifer Keffer
  Elizabeth Lentz

Traverse City Central High School
Traverse City, Michigan
  Dennis Muth
  Tonya Rice

Traverse City West High School
Traverse City, Michigan
  Tamie Rosenburg
  Diana Lyon-Schumacher
  John Sivek

Ypsilanti High School
Ypsilanti, Michigan
  Steve Gregory
  Mark McClure
  Beth Welch

Overview of Course 4

Part A

Unit 1 ▶ Rates of Change

Rates of Change develops student understanding of the fundamental concepts underlying calculus and their applications.

Topics include average and instantaneous rates of change, derivative at a point and derivative functions, accumulation of continuously varying quantities by estimation, the definite integral, and intuitive development of the fundamental theorem of calculus.

Lesson 1 *Instantaneous Rates of Change*
Lesson 2 *Rates of Change for Familiar Functions*
Lesson 3 *Accumulation at Variable Rates*
Lesson 4 *Looking Back*

Unit 2 ▶ Modeling Motion

Modeling Motion develops student understanding of two-dimensional vectors and their use in modeling linear, circular, and other nonlinear motion.

Topics include concept of vector as a mathematical object used to model situations defined by magnitude and direction; equality of vectors, scalar multiples, opposite vectors, sum and difference vectors, position vectors and coordinates; and parametric equations for motion along a line and for motion of projectiles and objects in circular and elliptical orbits.

Lesson 1 *Modeling Linear Motion*
Lesson 2 *Simulating Linear and Nonlinear Motion*
Lesson 3 *Looking Back*

Unit 3 ▶ Logarithmic Functions and Data Models

Logarithmic Functions and Data Models develops student understanding of logarithmic functions and their use in modeling and analyzing problem situations and data patterns.

Topics include inverses of functions; logarithmic functions and their relation to exponential functions, properties of logarithms, equation solving with logarithms; logarithmic scales and re-expression, linearizing data, and fitting models using log and log-log transformations.

Lesson 1 *Inverses of Functions*
Lesson 2 *Logarithmic Functions*
Lesson 3 *Linearizing Data*
Lesson 4 *Looking Back*

Unit 4 ▶ Counting Models

Counting Models extends student ability to count systematically and solve enumeration problems, and develops understanding of, and ability to do, proof by mathematical induction.

Topics include systematic counting, the Multiplication Principle of Counting, combinations, permutations; the Binomial Theorem, Pascal's triangle, combinatorial reasoning; the General Multiplication Rule for Probability; and the Principle of Mathematical Induction.

Lesson 1 *Methods of Counting*
Lesson 2 *Counting Throughout Mathematics*
Lesson 3 *The Principle of Mathematical Induction*
Lesson 4 *Looking Back*

Overview of Course 4

Part A (continued)

Unit 5 ▶ Binomial Distributions and Statistical Inference

Binomial Distributions and Statistical Inference extends student understanding of the binomial distribution, including its exact construction and how the normal approximation to the binomial distribution is used in statistical inference to test a single proportion and to compare two treatments in an experiment.

Topics include binomial probability formula; shape, mean, and standard deviation of a binomial distribution; normal approximation to a binomial distribution; hypothesis test for a proportion; design of an experiment; randomization test; and hypothesis test for the difference of two proportions.

Part B

Unit 6 ▶ Polynomial and Rational Functions

Polynomial and Rational Functions extends student ability to use polynomial and rational functions to represent and solve problems from real-world situations while focusing on symbolic and graphic patterns.

Topics include factored and expanded symbolic forms, computational complexity, connections between symbolic and graphical representations, multiplicity of zeroes, end behavior; Factor Theorem, Remainder Theorem, complex numbers and their use in the solution of polynomial equations, Fundamental Theorem of Algebra; equivalent forms of rational expressions; horizontal, vertical, and oblique asymptotes; and optimization.

Unit 7 ▶ Functions and Symbolic Reasoning

Functions and Symbolic Reasoning extends student ability to manipulate symbolic representations of exponential, logarithmic, and trigonometric functions; to solve exponential and logarithmic equations; to prove or disprove that two trigonometric expressions are identical and to solve trigonometric equations; to reason with complex numbers and complex number operations using geometric representations and to find roots of complex numbers.

Topics include equivalent forms of exponential expressions, definition of e and natural logarithms, solving equations using logarithms and solving logarithmic equations; the tangent, cotangent, secant, and cosecant functions; fundamental trigonometric identities, sum and difference identities, double-angle identities; solving trigonometric equations and expression of periodic solutions; rectangular and polar representations of complex numbers, absolute value, DeMoivre's Theorem, and the roots of a complex number.

Overview of Course 4

Part B (continued)

Unit 8  Space Geometry

Space Geometry extends student ability to visualize and represent non-regular three-dimensional shapes using contours, cross sections and reliefs and to visualize and represent surfaces and conic sections defined by algebraic equations.

Topics include using contours to represent three-dimensional surfaces and developing contour maps from data; sketching surfaces from sets of cross sections; conics as planar sections of right circular cones and as locus of points in a plane; three-dimensional rectangular coordinate system; sketching surfaces using traces, intercepts and cross sections derived from algebraically-defined surfaces; surfaces of revolution and cylindrical surfaces.

Lesson 1 *Representing Three-Dimensional Shapes*
Lesson 2 *Equations for Surfaces*
Lesson 3 *Looking Back*

Unit 9 ▶ Informatics

Informatics develops student understanding of the mathematics of information processing, focusing on the basic issues of access, security, and accuracy.

Topics include set theory; modular arithmetic; symmetric-key and public-key cryptosystems; error-detecting codes, including bar codes and check digits.

Lesson 1 *Access—The Mathematics of Internet Search Engines*
Lesson 2 *Security—Cryptography*
Lesson 3 *Accuracy—ID Numbers and Error-Detecting Codes*
Lesson 4 *Looking Back*

Unit 10 ▶ Problem Solving, Algorithms, and Spreadsheets

Problem Solving, Algorithms, and Spreadsheets develops student understanding and skill in use of standard spreadsheet operations for mathematical problems, while at the same time reviewing and extending many of the basic topics in Courses 1–3.

Topics include mathematics of finance, modeling population growth, apportionment of power in representative governments, sequences and series, and numerical solution of equations.

Lesson 1 *Money Adds Up and Multiplies*
Lesson 2 *Building a Library of Algorithms*
Lesson 3 *Mathematical Patterns in Shapes and Numbers*
Lesson 4 *Population Dynamics*
Lesson 5 *Population and Political Power*
Lesson 6 *Looking Back*

Contents

Preface

The first three courses in the *Contemporary Mathematics in Context* series provided a common core of broadly useful mathematics for all students. They were developed to prepare students for success in college, in careers, and in daily life in contemporary society. Course 4 continues the preparation of students for college mathematics. Formal and symbolic reasoning strategies, the hallmarks of advanced mathematics, are developed here as complements to more intuitive arguments and numerical and graphical approaches to problems developed in Courses 1–3.

Course 4 of the *Contemporary Mathematics in Context* curriculum shares many of the mathematical and instructional features of Courses 1–3.

- **Unified Content** Course 4 continues to advance students' mathematical thinking along interwoven strands of algebra and functions, statistics and probability, geometry and trigonometry, and discrete mathematics. These strands are unified by fundamental themes, by common topics, and by mathematical habits of mind or ways of thinking.

- **Mathematical Modeling** The curriculum emphasizes mathematical modeling including the processes of data collection, representation, interpretation, prediction, and simulation. Models developed in Course 4 come from many diverse areas including physics, economics, navigation, sports, health care, finance, biology, information processing, political science, sociology, and engineering.

- **Technology** Numerical, graphics, and programming/link capabilities such as those found on many graphing calculators are assumed and appropriately used throughout the curriculum. This use of technology permits the curriculum and instruction to emphasize multiple representations (verbal, numerical, graphical, and symbolic) and their use in modeling mathematical situations. Course 4 also introduces the use of spread-sheets as a problem-solving tool and the Internet as a source of rich applications.

- **Active Learning** Instructional materials promote active learning and teaching centered around collaborative small-group investigations of problem situations followed by teacher-led whole class summarizing activities that lead to analysis, abstraction, and further application of underlying mathematical ideas. Students are actively engaged in exploring, conjecturing, verifying, generalizing, applying, proving, evaluating, and communicating mathematical ideas.

- **Flexibility** The mathematical content and sequence of units in Course 4 allows considerable flexibility in tailoring a course to best prepare students for various undergraduate programs. For students intending to pursue programs in the *mathematical, physical, and biological sciences or engineering,* the developers recommend the following sequence of units:

  Unit 1 → Unit 2 → Unit 3 → Unit 4 →
  Unit 6 → Unit 7 → Unit 8 → Unit 5, 9, or 10

  For students intending to pursue programs in the *social, management, and some of the health sciences or humanities,* the following sequence of units is recommended:

  Unit 1 → Unit 2 (reduced) → Unit 3 →
  Unit 4 (reduced) → Unit 5 → Unit 9 → Unit 10

  The accompanying *Teacher's Guide* provides suggestions on how particular lessons in selected units can be omitted or streamlined without loss of continuity. Depending on time available, additional units of study can be selected based on student performance and interests.

- **Multi-dimensional Assessment** Comprehensive assessment of student understanding and progress through both curriculum-embedded assessment opportunities and supplementary assessment tasks supports instruction and enables monitoring and evaluation of each student's performance in terms of mathematical processes, content, and dispositions.

Unified Mathematics

Contemporary Mathematics in Context, Course 4 formalizes and extends important mathematical ideas drawn from four strands, with a focus on the mathematics needed to be successful in college mathematics and statistics courses.

The Algebra and Functions strand develops student ability to recognize, represent, and solve problems involving relations among quantitative variables. Central to the development is the use of functions as mathematical models. In Course 4, students extend their toolkit of function models to include logarithmic functions, polynomial functions, and rational functions. Function families are revisited in terms of the fundamental ideas of rates of change and accumulation. Increased attention is given to analysis of symbolic representations of functions. Students extend their skills in *symbolic manipulation*—rewriting expressions in equivalent forms, often to solve equations—and in *symbolic reasoning*—making inferences about symbolic relations and connections between symbolic representations and graphical, numerical, and contextual representations.

The primary goal of the Geometry and Trigonometry strand is to develop visual thinking and the ability to construct, reason with, interpret, and apply mathematical models of patterns in visual and physical contexts. In Course 4, concepts and methods of algebra, geometry, and trigonometry become increasingly intertwined in the development of models for describing and analyzing motion in two-dimensional space and surfaces in three-dimensional space.

The primary role of the Statistics and Probability strand is to develop student ability to analyze data intelligently, to recognize and measure variation, and to understand the patterns that underlie probabilistic situations. Graphical methods of data analysis, simulations, sampling, and experience with the collection and interpretation of real data are featured. In Course 4, ideas of probability distributions and data analysis are merged in the development of methods for testing a hypothesis. Work in the strand concludes with the design of experiments to produce data from which reliable conclusions can be drawn.

The Discrete Mathematics strand develops student ability to model and solve problems involving enumeration, sequential change, decision-making in finite settings, and relationships among a finite number of elements. Key themes are existence (Is there a solution?), optimization (What is the best solution?), and algorithmic problem-solving (Can you efficiently construct a solution?). A fourth theme introduced in Course 4 is that of proof, and in particular proof by mathematical induction. Abstract thinking required to construct proofs in discrete settings is also capitalized on in the development of combinatorial techniques that augment informal methods of systematic counting developed in prior courses. An introduction to the mathematics of information processing concludes work in this strand.

These four strands are connected within units by fundamental ideas such as symmetry, recursion, functions, re-expression, and data analysis and curve-fitting. The strands also are connected across units by mathematical habits of mind, such as visual thinking, recursive thinking, searching for and explaining patterns, making and checking conjectures, reasoning with multiple representations, inventing mathematics, and providing convincing arguments and proofs. The strands are unified further by the fundamental themes of data, representation, shape, and change. Important mathematical ideas are frequently revisited through this attention to connections within and across strands, enabling students to develop a robust and connected understanding of mathematics.

Active Learning and Teaching

The manner in which students encounter mathematical ideas can contribute significantly to the quality of their learning and the depth of their understanding. *Contemporary Mathematics in Context* Course 4 units are designed around multi-day lessons centered on big ideas. Lessons are organized around a four-phase cycle of classroom activities, described in the following paragraph. This cycle is designed to engage students in investigating and making sense of problem situations, in constructing important mathematical concepts and methods, in generalizing and proving mathematical relationships, and in communicating, both orally and in writing, their thinking and the results of their efforts. Most classroom activities are designed to be completed by students working collaboratively in groups of two to four students.

The launch phase promotes a teacher-led class discussion of a problem situation and of related questions to think about, setting the context for the student work to follow. In the second or explore phase, students investigate more focused problems and questions related to the launch situation. This investigative work is followed by a teacher-led class discussion in which students summarize mathematical ideas developed in their groups, providing an opportunity to construct a shared understanding of important concepts, methods, and approaches. Finally, students are given a task to complete on their own, assessing their initial understanding of the concepts and methods.

Each lesson also includes tasks to engage students in Modeling with, Organizing, Reflecting on, and Extending their mathematical understanding. These MORE tasks are central to the learning goals of each lesson and are intended primarily for individual work outside of class. Selection of tasks for use with a class should be based on student performance and the availability of time and technology. Students can exercise some choice of tasks to pursue, and at times they can be given the opportunity to pose their own problems and questions to investigate.

Following each MORE set, there is a Preparing for Undergraduate Mathematics Placement (PUMP) exercise set providing practice in skills and reasoning techniques commonly assessed on college mathematics placement tests.

Multiple Approaches to Assessment

Assessing what students know and are able to do is an integral part of *Contemporary Mathematics in Context*. Initially, as students pursue the investigations that make up the curriculum, the teacher is able to informally assess student understanding of mathematical processes and content and their disposition toward mathematics. At the end of each investigation, the Checkpoint and accompanying class discussion provide an opportunity for the teacher to assess the levels of understanding that various groups of students have reached as they share and summarize their findings. Finally, the "On Your Own" problems, the tasks in the MORE sets, and the exercises in the PUMP sections provide further opportunities to assess the level of understanding of each individual student. Quizzes, in-class exams, take-home assessment tasks, and extended projects are included in the teacher resource materials.

Acknowledgments

Development and evaluation of the student text materials, teacher materials, assessments, and calculator software for Course 4 of *Contemporary Mathematics in Context* was funded through a grant from the National Science Foundation to the Core-Plus Mathematics Project (CPMP). We express our appreciation to NSF and, in particular, to our program officer John Bradley for his continuing trust, support, and input.

We also are grateful to Texas Instruments and, in particular, Dave Santucci, for collaborating with us by providing classroom sets of graphing calculators to field-test schools.

As seen on page vii, CPMP has been a collaborative effort that has drawn on the talents and energies of teams of mathematics educators at several institutions. This diversity of experiences and ideas has been a particular strength of the project. Special thanks is owed to the exceptionally capable support staff at these institutions, particularly at Western Michigan University.

We are also grateful to our Advisory Board, Diane Briars (Pittsburgh Public Schools), Gail Burrill (University of Wisconsin–Madison), Jeremy Kilpatrick (University of Georgia), Kenneth Ruthven (University of Cambridge), David A. Smith (Duke University), and Edna Vasquez (Detroit Renaissance High School) for their ongoing guidance and advice.

The overall design and mathematical focus of Course 4 was informed by a working conference with prominent mathematicians held at the University of Maryland in September 1996. Participants included Jim Lewis (University of Nebraska–Lincoln), Steve Rodi (Austin Community College), Kenneth Ross (University of Oregon), Richard Scheaffer (University of Florida), Al Taylor (University of Michigan–Ann Arbor), Tom Tucker (Colgate University), and Paul Zorn (St. Olaf College). We greatly appreciate their insights on what students should know and be able to do upon entering college. Special thanks are owed to Kenneth Ross, Richard Scheaffer, and Paul Zorn who, in addition, reviewed and commented on units as they were being developed, tested, and revised.

Our gratitude is expressed to the teachers and students in our 32 evaluation sites listed on pages viii and ix. Their experiences using pilot- and field-test versions of *Contemporary Mathematics in Context* Course 4 provided constructive feedback and improvements. We learned a lot together about making mathematics meaningful and accessible to a wide range of college-bound students.

A very special thank you is extended to Anna Belluomini for her interest and support in publishing a fourth-year college preparatory course that breaks new ground in terms of content, instructional practices, and student assessment. Finally, we want to acknowledge Abby Tanenbaum and Luke Zajac for their thoughtful and careful editorial work and express our appreciation to the editorial staff of Glencoe/McGraw-Hill who contributed to the publication of this program.

To the Student

Contemporary Mathematics in Context, Course 4 builds on the mathematical concepts, methods, and habits of mind developed in Courses 1–3 with particular attention to building a solid bridge to collegiate mathematics. A major focus of this course is the development of formal and symbolic reasoning strategies as a complement to more intuitive arguments and numerical and graphical approaches to problems you used in previous courses.

With this text, you will continue to learn mathematics by doing mathematics, not by memorizing "worked out" examples. You will investigate important mathematical ideas and ways of thinking as you try to understand and make sense of realistic situations. Because real-world situations and problems often involve data, shape, change, or chance, you will extend and formalize your understanding of fundamental concepts and methods from algebra and functions, from statistics and probability, from geometry and trigonometry, and from discrete mathematics. You also will see further connections among these strands—how they weave together to form the fabric of mathematics.

Because real-world situations and problems are often open-ended, you will find that there may be more than one correct approach and more than one correct solution. Therefore, you will frequently be asked to explain your ideas. You also will increasingly be asked to provide more formal arguments or proofs for mathematical statements. This text will provide help and practice in reasoning and communicating clearly about mathematics.

Because the solution of real-world problems often involves teamwork, you will continue to often work collaboratively with a partner or in small groups as you investigate realistic and interesting situations. As in Courses 1–3, you will find that 2 to 4 students working collaboratively on a problem can often accomplish more than any one of you would working individually. Because technology is commonly used in solving real-world problems, you will continue to use a graphing calculator or computer as a tool to help you understand and make sense of situations and problems you encounter.

Most colleges and universities administer a mathematics placement test to incoming students. The results of this test, together with grades in high school mathematics courses and intended undergraduate major, determine the first college mathematics course you will study. To perform well on these tests, there are a number of skills and reasoning techniques that need to be automatic. Developing that level of proficiency requires practice. To help you strengthen your skills in strategic areas, we have included special Preparing for Undergraduate Mathematics Placement (PUMP) exercise sets in each unit.

As in Courses 1–3, you will continue to learn a lot of useful mathematics, and it is going to make sense to you. You also will deepen your understanding of fundamental ideas that support future coursework in mathematics and statistics. You are going to strengthen your skills in working cooperatively and communicating with others as well. You are also going to strengthen your skills in using technological tools intelligently and effectively. Finally, you will continue to have plenty of opportunities to be creative and inventive. Enjoy.

Polynomial and Rational Functions

Unit 6

Polynomial Functions

The movie *Toy Story* was the first feature film animated entirely by computer. It took 27 animators, 800,000 machine hours and 160 billion pixels to create the look and feel of *Toy Story*. The storage capacity of 1,000 CDs (650,000 megabytes) was required to hold all of the film's data.

© Disney Enterprises, Inc.

Now, more than ever, computers and mathematics are used in special effects for movies, animation, and product design. Most of these tasks require the ability to work with smooth curves quickly and easily. The curves are often specified by giving key *control points* on the curves.

Think About This Situation

Examine this model of the left forearm of Buzz Lightyear. Think about creating a curve similar to a portion of his forearm through the indicated control points.

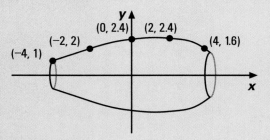

ⓐ What strategies can you use to find an equation whose graph goes through the five control points indicated?

ⓑ Could another curve be drawn which passes through the five points? Explain your reasoning.

ⓒ What characteristics of a curve might determine how many control points are necessary to determine its shape?

ⓓ How might control points be useful in animating the movements of Buzz's arm?

INVESTIGATION 1 ▶ Starting at Zeroes

Being able to specify a curve by control points is an important part of almost all computer graphics design and animation. When trying to find a function model for a curve, the simplest of all cases occurs when the control points correspond to the zeroes of a *polynomial function.*

In previous units, you saw that the height h (in feet) of an object thrown or kicked up in the air can be modeled by a quadratic function $h(t) = -16t^2 + bt + c$, where t is measured in seconds. For example, the height of a soccer ball kicked with an initial upward velocity of 50 feet per second can be modeled by the function $h(t) = 50t - 16t^2$. In your earlier work, you discovered that a quadratic function can sometimes be written in the factored form $h(t) = k(t - a)(t - b)$, where a and b are the zeroes of the function and k is a constant. You may also recall that every quadratic function has 0, 1, or 2 real-number zeroes. These facts can help in modeling curves for motion.

1. In football, when a team kicks off to the opposing team, the *hang time* (time in the air) of a kickoff is sometimes as important as the distance of the kickoff.

   a. Suppose an announcer indicates that the hang time for a kickoff was 3.5 seconds. At what times can you assume that the height of the football was 0 ft?

   b. Write an equation in factored form that expresses the height h of the ball in feet as a function of elapsed time t.

   c. Now rewrite your function $h(t)$ in standard polynomial form so that the squared term has a coefficient of -16. Be sure that the zeroes remain the same.

   d. What information can you infer from the polynomial form of the height function?

   e. Based on your experiences, when should the football reach its maximum height? At what time does the football reach its maximum height? What is the maximum height?

   f. Why was knowing the hang time for a kickoff sufficient to determine a model for the height of the football as a function of time?

2. One way to approximate the depth of a well is to determine the time it takes for an object to drop from the top of the well to the surface of the water. To get a precise measurement, you would also have to take into account the influence of air friction, terminal velocity, and the speed of sound. Suppose you dropped a pebble into a well and 3 seconds passed before you heard a resounding "plop."

a. Sketch a graph of the height of the pebble above the surface of the water as a function of elapsed time. Label key points on your graph.

b. Explain why the graph is not a diagram of the path of the pebble.

c. Your graph for Part a should be half of a parabola. Add to your graph the other half of the parabola.

d. Describe a situation involving a well and throwing a pebble, where the graph of the height of the pebble above the surface of the water would match your graph in Part c.

e. Write an equation in factored form that expresses the height h of the pebble above the surface of the water as a function of elapsed time t.

f. How deep is the well?

In Activities 1 and 2, knowing the type of function that would model each situation and its zeroes was helpful in determining the quadratic function in each case. The same is true for more complicated curves modeled by higher-degree polynomial functions. **Polynomial functions** are functions that can be expressed in the following form:

$$P(x) = a_n x^n + a_{n-1} x^{n-1} + a_{n-2} x^{n-2} + \ldots + a_1 x + a_0,$$
where n is an integer, $n \geq 0$

The highest power of the polynomial is called the **degree** of the polynomial, each member of the sum, $a_i x^i$, is called a **term**, and the constants a_i are called **coefficients**. For example, $P(x) = 5x^4 + 10x^3 - 65x^2 - 70x + 120$ is a polynomial function of degree four with five terms having coefficients 5, 10, −65, −70, and 120, respectively.

The polynomial function $P(x) = 5x^4 + 10x^3 - 65x^2 - 70x + 120$ can be rewritten in *factored form* as $P(x) = 5(x - 1)(x + 2)(x - 3)(x + 4)$, with zeroes of 1, −2, 3, and −4. Each expression of the form $(x - a)$ or $(x + a)$ is called a **linear factor** of the polynomial $P(x)$.

As in the case of quadratic functions, not all polynomial function rules can be factored into linear factors. For example, the polynomial function $P(x) = x^4 + 1$ cannot be expressed as a product of linear factors using real numbers.

In the next two activities, you will explore the usefulness of polynomial functions of degree greater than two.

3. Highway construction engineers are frequently faced with the problem of approximating regions of land in order to predict how much dirt must be moved, how much concrete is needed, how much landscaping will be required, and so on. To do this, they fit low-degree polynomial curves to the planned shape.

In the following diagram, an S-shaped curve is being planned between the three points. In practice, civil engineers would find curves for both edges of the road. To simplify the problem, just consider the general shape of the curve.

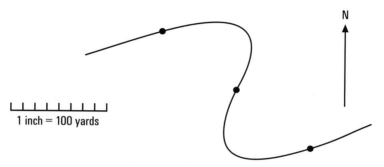

1 inch = 100 yards

a. Carefully trace the diagram on a sheet of paper. Consider possible placements of coordinate axes to aid in modeling this situation.

b. Explain how placing one of the axes through the three points would help determine an equation of the curve. Where would you place the other axis? Use this placement of axes and the given scale to measure important distances.

c. Find a polynomial function expressed in factored form whose graph resembles the shape of the curve. What is the degree of your polynomial?

d. How well does your function predict the location of points along the curve? How could you modify the function rule to improve the fit? Compare your modified rule to those of other groups.

e. How do you think your modified function might be used in the planning or construction of the road?

4. Roller coasters are designed and tested on computers before they are built. Suppose an engineer wishes to model a small section of track that resembles a W-shaped curve. In practice, the 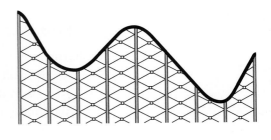 engineer might use several polynomial functions pieced together to design such a track. Consider a fourth-degree or *quartic polynomial* function to model the curve. Suppose the axes are placed as shown in the diagram below.

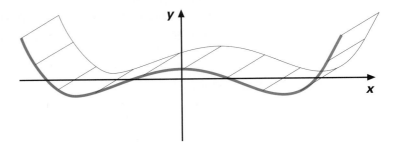

a. If the curve crosses the *x*-axis at $x = -6, -2, 2,$ and 6 meters, what is a possible fourth-degree polynomial function that might model this curve?

- How do you know your polynomial is of degree 4?

- How would you restrict the domain of the function so that its graph resembles the section of the track shown?

b. The locations of the lowest points of the track are approximately at $(-4.5, -1)$ and $(4.5, -1)$. Adjust your polynomial function so that these points are the lowest points on its graph.

c. Rather than using a curve where the height of the track can be negative, an engineer is more likely to place the horizontal axis so that it contains the lowest points along the track. How would you transform the given graph and function in Part b so that the lowest points, which occur at $x = -4.5$ and $x = 4.5$, are on the *x*-axis?

d. Finally, adjust your function rule in Part c so that the restricted domain is all nonnegative numbers.

e. What other methods could you use to find an equation of the curve that satisfies the conditions in Parts c and d?

Checkpoint

Knowing where a curve intersects the *x*-axis or knowing the zeroes of a corresponding function is helpful in determining an equation for the curve.

a Write an equation of a polynomial function that has zeroes –3, –2, 0, 3, 4, and 10.

b Does knowing the zeroes of a polynomial function uniquely determine the function? Explain.

c If a polynomial function has eight distinct zeroes, then what is the minimum degree of the polynomial?

Be prepared to explain your equation and thinking to the entire class.

On Your Own

Graphs of quadratic polynomial functions have one basic shape (a parabola) that can be translated or scaled (stretched or compressed) to create a curve with 0, 1, or 2 real-number zeroes. However, for higher-degree polynomials, graphs of two polynomial functions of the same degree can have different shapes.

a. Find a *cubic* (degree 3) polynomial function whose graph matches that shown in each case.

i.
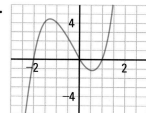

ii.
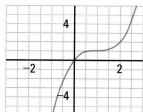

b. As each curve is shifted vertically, what happens to the number of zeroes of the corresponding function?

c. What are characteristics of the two graphs that justify saying their shapes are different?

d. Show that the graphs of the two quartic functions $y = x^4$ and $y = x^4 - 4x^3$ have different shapes, and that both functions can have either 0, 1, or 2 zeroes in their original positions or when shifted vertically.

INVESTIGATION 2 Getting to the Points

You can use the statistical regression feature of your calculator to fit a polynomial function of degree four or less to a set of points. Regression methods are particularly useful when the points represent measurement data and the desired graph does not have to pass precisely through the given points. However, with computer animation and computer-aided design, the points are predetermined exactly and the curves must pass through each point.

1. Consider the control points used at the beginning of this lesson to model Buzz Lightyear's left forearm.

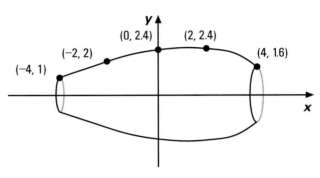

© Disney Enterprises, Inc.

a. Use the regression feature of your calculator to find a quadratic, a cubic, and a quartic polynomial function that best fits the pattern of the five points.

b. Which, if any, of your polynomial functions in Part a have graphs that go through all five points?

c. Thinking about how this curve might be used in combination with other curves to create the image of Buzz Lightyear, why is it important for the curve to contain the points?

Finding a polynomial function whose graph passes through a set of points is not too difficult when the points can be thought of as zeroes of the function. However, as in the case of Buzz's arm, control points are rarely a set of zeroes for a polynomial function. One general method of finding a polynomial function whose graph passes through a set of points is called the **method of undetermined coefficients**. Recall the general form of a polynomial function of degree n:

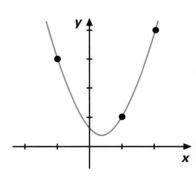

$$P(x) = a_n x^n + a_{n-1} x^{n-1} + \ldots + a_1 x + a_0$$

For a polynomial of degree n, there are $(n + 1)$ coefficients, $a_0, a_1, ..., a_n$ that must be determined. Points on the graph can be used to generate equations that the coefficients must satisfy. For example, suppose that one of the given points is $(2, 4)$. The desired polynomial function $P(x)$ must satisfy the equation $P(2) = 4$. Thus, the coefficients must satisfy the following equation:

$$a_n \cdot 2^n + a_{n-1} \cdot 2^{n-1} + ... + a_1 \cdot 2 + a_0 = 4$$

To determine $(n + 1)$ unknowns, $(n + 1)$ equations are needed. Therefore, a set of $(n + 1)$ points are required to determine a polynomial of degree n.

2. Assume that three points $(-1, 3)$, $(1, 1)$, and $(2, 4)$ determine a parabola with equation of the form $y = ax^2 + bx + c$, where a, b, and c are the undetermined coefficients.

   a. Explain why each of the following equations must be true:

   $$3 = a - b + c$$
   $$1 = a + b + c$$
   $$4 = 4a + 2b + c$$

   b. How can the first two equations be used to solve for one of the coefficients a, b, or c?

   c. Solve for the remaining two coefficients. Write the equation of the parabola.

   d. Describe three different methods that you could use to verify your equation. Use one of these methods.

   e. Use the method of undetermined coefficients to find a cubic polynomial function whose graph contains the four points $(-2, -3)$, $(-1, 8)$, $(0, 7)$, and $(2, 17)$.

Using the method of undetermined coefficients leads to a system of linear equations. In your previous work, you have solved systems of linear equations using matrices.

3. Refer back to Part a of Activity 2.

   a. Explain why the following matrix equation is true.

   $$\begin{bmatrix} 1 & -1 & 1 \\ 1 & 1 & 1 \\ 4 & 2 & 1 \end{bmatrix} \begin{bmatrix} a \\ b \\ c \end{bmatrix} = \begin{bmatrix} 3 \\ 1 \\ 4 \end{bmatrix}$$

   b. Why is the following equation true?

   $$\begin{bmatrix} a \\ b \\ c \end{bmatrix} = \begin{bmatrix} 1 & -1 & 1 \\ 1 & 1 & 1 \\ 4 & 2 & 1 \end{bmatrix}^{-1} \begin{bmatrix} 3 \\ 1 \\ 4 \end{bmatrix}$$

   c. Using the matrix capabilities of your calculator, determine values for a, b, and c. Compare your polynomial function to the one found in Activity 2.

   d. Extend this matrix method to find the cubic polynomial function whose graph passes through the points given in Part e of Activity 2.

In this investigation, you explored the method of undetermined coefficients for finding a polynomial function whose graph passes through a given set of points.

ⓐ What other methods do you know for finding a polynomial function whose graph passes through a given set of points?

ⓑ Use the method of undetermined coefficients to find a polynomial function whose graph contains the points $(1, 2)$, $(3, 4)$, and $(5, 0)$.

ⓒ If you were to use the method of undetermined coefficients to find a polynomial function whose graph passes through five given points, what would be the largest possible degree of the polynomial? Explain your reasoning.

Be prepared to report on your understanding of methods for fitting polynomials to sets of points.

On Your Own

Two points in a coordinate plane determine a straight line. Similarly, as you saw in this investigation, three noncollinear points uniquely determine a parabola.

a. Using the method of undetermined coefficients, find the equation of the parabola that passes through the points $(1, -4)$, $(3, 0)$, and $(4, 1)$.

b. Use another method to find the quadratic equation in Part a.

INVESTIGATION 3 ▶ A Matter of Degree

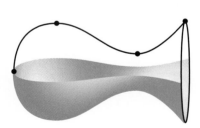

In designing curves, once a set of control points has been fixed, you can use the method of Investigation 2 to find a polynomial function whose graph passes through the set of points. The maximum degree of the polynomial is determined by the number of points. But how is the number of points chosen? Often the selection is made by comparing characteristics of the curve with characteristics of graphs of polynomial functions of different degrees. In this investigation, you will explore classification characteristics of polynomial functions.

1. One useful classification characteristic is called the **end behavior** of a function, which is the behavior of the function for positive and negative values of x with very large absolute values.

   a. Look more closely at the end behavior of quadratic functions.

   ■ What happens to the function values of $f(x) = 2x^2 - x - 10$ as the x values get larger and larger or as the *x values approach positive infinity* (denoted $x \to +\infty$)? As the x values approach negative infinity ($x \to -\infty$)?

   ■ Describe the end behavior of the quadratic function $g(x) = -3x^2 + 27x - 1{,}000$.

■ What term of a quadratic polynomial function rule can be used to predict the end behavior of the function? Why does your observation make sense?

b. Now explore the end behavior of cubic polynomial functions and summarize your findings. What term of a cubic polynomial function rule can be used to predict the end behavior of the function?

c. How is the end behavior of a quadratic polynomial function different from the end behavior of a cubic polynomial function?

d. Explain why the graph of any polynomial function, when viewed on a large enough scale, will look like the graph of a power function.

2. Another important characteristic of polynomial functions is the number of *peaks* or *valleys* of their graphs. A peak is called a **local maximum** (plural: maxima) and a valley is called a **local minimum** (plural: minima).

a. What is the total number of local maxima and minima that a quadratic polynomial function can have? Why?

b. Shown below are the graphs of two polynomial functions $y = g(x)$ and $y = h(x)$. The scale on each axis is 1.

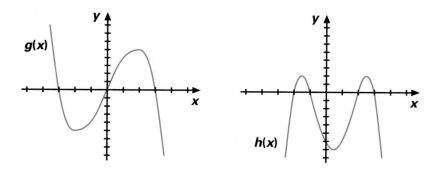

■ Identify the local maxima and minima of $y = g(x)$.

■ Identify the local maxima and minima of $y = h(x)$.

c. Give an example of a polynomial function with no local maximum or minimum.

d. Explain why the adjective "local" is used in describing peaks and valleys of polynomial functions.

e. A polynomial function $f(x)$ can have an absolute maximum or an absolute minimum as well. An **absolute maximum** is a value M for which $f(x) \leq M$ for all values of x. An **absolute minimum** is a value m for which $f(x) \geq m$ for all values of x.

■ Explain why the function $g(x)$ in Part b does not have an absolute maximum or an absolute minimum.

■ Does the function $h(x)$ in Part b have an absolute maximum or an absolute minimum? Explain.

f. Look back at the graphs of the polynomial functions in Part b.

- Find an equation for $g(x)$ if $g(1) = 4$.
- Find an equation for $h(x)$.

3. In this activity, you will explore characteristics of cubic polynomial functions of the form $f(x) = ax^3 + bx^2 + cx + d$.

a. Explain why at least four points are necessary to determine a cubic polynomial function.

b. Using different sets of four points, create as many different shapes as you can that would be possible graphs of cubic polynomial functions.

- Record a sample function for each different shape.
- Use both small and large viewing windows to make sure you are seeing the complete behavior of the function.
- Indicate what characteristics you are using to categorize the shapes of the graphs.

c. List the possible numbers of local maxima and minima of a cubic polynomial function.

d. List the possible numbers of zeroes of a cubic polynomial function.

An important characteristic that is used to describe a polynomial function is the number of zeroes. For example, the polynomial function $y = (x - 1)^3 = (x - 1)(x - 1)(x - 1)$ is said to have a **zero of multiplicity 3** at $x = 1$. As another example, the polynomial function $y = (x + 2)(x + 2)(x - 3) = (x + 2)^2(x - 3)$ has one zero, $x = -2$, of multiplicity 2 and one zero, $x = 3$, of multiplicity 1.

4. The shape of the graph of a polynomial function near a zero provides information about the possible multiplicity of the zero.

a. Graph the polynomial function $f(x) = (x - 3)^n$ for different positive integer values of n. Examine the behavior of each graph near $x = 3$. What appears to be true about the behavior of a polynomial function near a zero of multiplicity n?

b. Test your conjecture in Part a by considering the zeroes of the polynomial function $g(x) = (x + 3)^2(x + 1)^3(x - 1)$.

c. Examine the graph of $y = h(x)$ below. The scale on both axes is 1. Find an equation for $h(x)$ if $h(-1) = -3$.

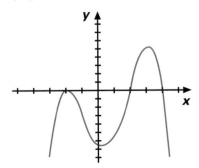

5. Summarize your findings from Activities 1–4 in a table similar to the one below. Sharing the workload with others in your group, repeat similar activities for polynomial functions of degree 4, 5, and 6. Record your findings in your table.

Properties of Polynomial Functions

| Characteristic | Polynomial Degree | | | | |
	2	3	4	5	6
Number of Points Needed to Determine the Polynomial					
Possible Shapes					
Number of Different Shapes					
Possible Number of Local Maxima and Minima					
Absolute Maximum or Minimum?					
Possible Numbers of Zeroes Counting Multiplicity					
Possible Numbers of Zeroes Ignoring Multiplicity					
Behavior as $x \rightarrow \pm\infty$					
Other Observations					

a. What patterns do you see in the table based upon the degree of the polynomial? Compare your observations to those of other groups. Resolve any differences.

b. Test your patterns by making and checking conjectures about the characteristics of polynomial functions of degree 7.

c. Suppose $y = f(x)$ is a polynomial function of degree n, $n \geq 2$. What can you conclude about the characteristics of this function? Explain your reasoning.

Checkpoint

In this investigation, you examined several key characteristics of polynomial functions.

ⓐ What are the possible combinations of local maxima and local minima for a sixth-degree polynomial function? Explain your reasoning.

ⓑ If a polynomial function has an absolute maximum or an absolute minimum, what can you say about the degree of the polynomial?

ⓒ How do the graphs of polynomial functions of even and odd degree differ?

ⓓ How many points are needed to approximate the following curve with a polynomial function? Explain your reasoning.

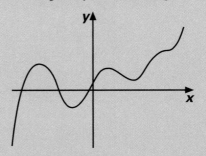

Be prepared to compare your responses to those of other groups.

▶On Your Own

A polynomial function *f* has exactly three zeroes: −3, 1, and 4, each with multiplicity 1, and $f(3) = -4$.

a. What can you conclude about the local maxima/minima of *f*?

b. What can you conclude about the absolute maximum/minimum of *f*?

c. Write an equation for $f(x)$.

d. Describe the end behavior of $y = f(x)$.

MORE

Modeling • Organizing • Reflecting • Extending

Modeling

1. Consider the following open-top box layout with cuts (solid lines) and folds (dotted lines) as indicated. The box is made by cutting out the corners but leaving sufficient material as flaps for gluing or tucking in.

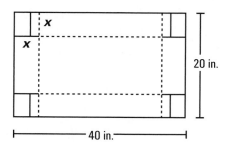
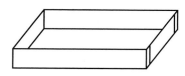

a. Express the volume V of the resulting box as a function of the length of the corner cutout, x.

- What is the degree of your polynomial function?
- What is the practical domain for this function?
- What are the zeroes of this function?

b. Draw a sketch of the graph of $y = V(x)$. Explain why there must be a value of x that maximizes the volume of the box.

c. Determine the maximum volume.

2. Refer to Modeling Task 1 above. Let a be the width and b be the length of the cardboard instead of the specific values given in the figure.

a. Write an equation for the volume of the box. What are the theoretical and practical domains of your equation for volume?

b. Draw a sketch of the graph of your equation, assuming $a < b$.

c. Based on the equation and your sketch, explain why, for any values of a and b, there will be a value of x which maximizes the volume. Indicate in what range of x values the maximum must occur.

d. Now assume that the piece of cardboard is twice as long as it is wide, that is, $b = 2a$.

- Create a table that gives the optimal volume and corresponding value of x for different values of a.
- Find a model relating the value of a and the corresponding optimal value of x.

3. Suppose you wish to construct a box where the top is attached or "hinged" on one side to the bottom as shown at the right.

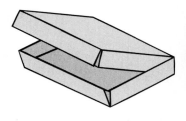

 a. Design a template showing the cuts and folds necessary. Write a formula for the volume of the box if it is made from a sheet of 40 in. × 20 in. cardboard and the length of the square corner cutout is x.

 b. What dimensions for the box will give the maximum volume? What is the volume of this box?

4. The shaft of a turbine engine design was created using the five indicated control points. Think about creating a curve to serve as the basis for the turbine shaft shown.

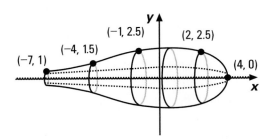

 a. Based on the shape of the curve, what is the minimum number of points necessary to get the general shape of the shaft?

 b. Why do you think five points were used?

 c. Find a polynomial function whose graph passes through the five points.

 d. How might the curve be used to create the solid shape of the shaft?

Organizing

1. The characteristics of polynomial functions you discovered in Investigation 1 are helpful in sketching graphs and in judging the reasonableness of calculator- or computer-produced graphs. Testing the sign of a polynomial expression is another useful aid in sketching graphs. Study the *sign diagram* below for the polynomial function $f(x) = x(x - 2)(x + 2)$.

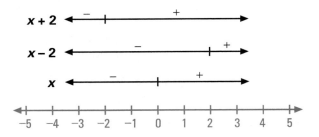

a. Use the sign diagram to determine any intervals in which $f(x) > 0$ and any intervals in which $f(x) < 0$.

b. Use the information derived from the sign diagram to make a rough sketch of $y = f(x)$.

c. Make a rough sketch of $g(x) = (x - 4)(x - 1)(x + 3)$ by reasoning with the sign of each factor.

2. The primary use of linear factors in Investigation 1 was to create polynomial functions with a given set of zeroes. Graphs associated with the factors of a polynomial function can also provide insight into the graph of the polynomial function more globally.

a. Graph the two linear functions $Y_1 = x - 3$ and $Y_2 = x + 4$, together with the function $Y_3 = Y_1 \cdot Y_2$. Relate the three graphs by discussing the location of zeroes and the sign of each function in different intervals.

b. If the function Y_1 in Part a is changed to $Y_1 = 3 - x$, will the same relationship exist among Y_1, Y_2, and Y_3 as in Part a?

c. Graph the three linear functions $Y_1 = x - 2$, $Y_2 = x$, and $Y_3 = x + 2$, together with the function $Y_4 = Y_1 \cdot Y_2 \cdot Y_3$. Discuss how the general shape of the graph of the product of the three functions can be determined based upon the graphs of the three linear functions.

d. As you have previously seen, not all polynomials can be written as a product of linear factors using real numbers. Graph the functions $Y_1 = x - 2$, $Y_2 = x^2 + 4$, and the product $Y_3 = Y_1 \cdot Y_2$. What is the degree of this product? Does the product function have the signs and zeroes that you expected?

e. Explain why the cubic function Y_3 from Part d cannot be written as a product of three linear factors.

3. You have previously seen that not every quadratic polynomial is factorable into polynomials of lower degree using real numbers. However, a group of students at Prairie High School claimed that every cubic polynomial can be factored into polynomials of degree 1 or 2. Their reasoning  is given below. Decide whether or not each statement is true and why.

i. If $f(x)$ is any cubic polynomial function, then $f(x)$ has at least one zero, say $x = a$.

ii. If a is a zero of $f(x)$, then $f(x) = (x - a)g(x)$.

iii. Since $g(x)$ is a polynomial function, $g(x)$ must be a polynomial of degree 2.

iv. Therefore, every cubic polynomial can be factored into polynomials of degree 1 or 2.

4. A little information about the graph of a cubic polynomial function can yield further insight about the shape of the graph and the nature of the symbolic rule for the function.

   a. Suppose you know that the graph of a cubic polynomial function intersects the x-axis at $x = 1$, $x = 2$, and $x = 3$. What do you know about the shape of the graph and a possible equation for the function? What additional information would allow you to write the equation for the function?

   b. Suppose you know that the graph of a specific cubic polynomial function intersects the x-axis at least twice. What can you say about the shape of the graph of the function? Explain.

   c. What can you say about the equation for a cubic polynomial function whose graph intersects the x-axis exactly twice?

5. A polynomial function is a power function or the sum of two or more power functions, each of which has a nonnegative integer power. In Unit 1, "Rates of Change," you may have discovered that the derivative function for a power function of degree n is a power function of degree $n - 1$.

   a. Graph the function $f(x) = x^3 - 6x^2 + 9x$ and an approximate derivative function $d(x) = \frac{f(x + 0.001) - f(x)}{0.001}$.

   b. What are the zeroes of the function $d(x)$? Explain why the zeroes of the derivative occur where $f(x)$ has a local maximum or local minimum.

   c. Using three points on the graph of $d(x)$, find a quadratic equation that matches the graph of $d(x)$.

   d. How does the graph of your equation compare to the graph of the approximate derivative?

   e. How does your equation compare to the original function $f(x)$?

   f. Using your earlier conjecture about the maximum possible number of local maxima and minima, explain why a polynomial function of degree 6 will have a derivative function of degree at most 5.

Reflecting

1. Explain why the polynomial $0x^3 + 2x^2 - 5x + 7$ is said to be of degree 2 rather than of degree 3. What problems do you think might arise if this were not the case?

2. In Activity 2 of Investigation 1 (page 362), an experiment was described for measuring the depth of a well. It was stated that if a precise measurement of the depth was needed, then the influence of air friction, terminal velocity, and speed of sound would need to be taken into account. If these factors are not considered, would your estimate of the depth of the well be too high or too low? Why?

3. A basic geometric fact is that "two points determine a line." Under what conditions will two points determine a parabola?

4. In Investigation 2, you used two different methods to find the equation of a quadratic function whose graph passes through three given noncollinear points. What happens if each method is applied to three points which are collinear?

5. Explain how it is possible for a function to have a local maximum without having an absolute maximum.

Extending

1. Finding a polynomial function whose graph passes through a set of points cannot always be done or may not produce desired results. For example, curves such as those used in this computer-rendered airplane must pass through the control points in a smooth natural manner.

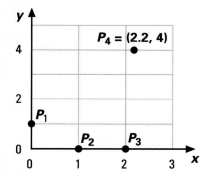

a. Sketch by hand a smooth curve that passes through the four points indicated in the graph above.

b. Where on the airplane shown might the curve you sketched by hand arise?

c. Find a function $y = ax^3 + bx^2 + cx + d$ whose graph passes through the four points.

d. How does the graph of the cubic polynomial function that passes through the four points compare with the curve you sketched by hand?

2. Consider the function $p(x) = 4x^3 - 6x^2 - 8x + 12 = ((4x - 6)x - 8)x + 12$.

   a. Without using a calculator, find $p(2)$ using each form of the polynomial. For which symbolic form was the calculation easiest?

   b. Examine this shorthand way of evaluating $p(2)$.

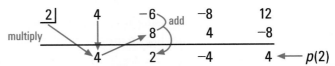

   How is the above procedure related to evaluating $p(2)$ using the *nested multiplication* form in Part a?

   c. The procedure outlined in Part b is called **synthetic substitution**. Use synthetic substitution to evaluate $f(x) = 5x^4 - 2x^3 + 10x^2 - x - 5$ for $x = 2$ and $x = -3$.

   d. Why do you think the procedure is called synthetic substitution?

3. Finding polynomial functions whose graphs contain a set of points can be combined with parametric equations to produce more complicated curves. Examine the loop in the roller coaster track shown below.

   a. What four control points do you think might produce a curve similar to the loop shown? Assign times starting at $t = 0$ for the roller coaster to be at each location. Create a table of values for (t, x, y).

   b. Using any method, find a polynomial function $P_1(t)$ for the points (t, x). Find a second polynomial function $P_2(t)$ for the points (t, y).

c. Use your calculator to plot the curve given parametrically by the following pair of equations:

$$x = P_1(t)$$

$$y = P_2(t)$$

Indicate an appropriate viewing window, including a range for t.

4. Apply the method of Extending Task 3 to the problem in Extending Task 1 using time values of $t = 1, 2, 3$, and 4 along the curve you drew by hand. Why do you think the curve created in this manner is smoother and more natural than the graph of the polynomial function found by using cubic regression?

5. Write and test a calculator or computer program that implements the method of undetermined coefficients for finding a polynomial function whose graph passes through a given set of points.

1. In a room with 300 people, there are 50% more men than women. How many women are in the room?

   (a) 60   (b) 120   (c) 180   (d) 150   (e) 100

2. Simplify: $\dfrac{(x+y)^2 - x^2}{y}$

   (a) $2xy + y$ (b) y (c) 0 (d) $2x + y$ (e) $2x + y^2$

3. Solve simultaneously for x and y: $2x + 4y = -6$, $x + 3y = -4$.

   (a) $(1, 1)$ (b) $(2, 2)$ (c) $(-1, 1)$ (d) $(-1, -1)$ (e) $(-3, -3)$

4. Which of the following could not be a portion of the graph of a function $y = f(x)$?

   (a) 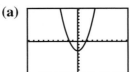   (b) 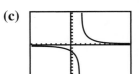   (c)

   (d) 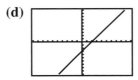   (e)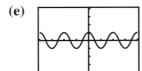

5. What is the maximum point of the graph of the function $f(x) = -(x - 5)^2 - 12$?

   (a) $(5, -12)$ (b) $(-5, -12)$ (c) $(-5, 12)$ (d) $(5, 12)$ (e) $(-12, -5)$

6. If $x \neq -3$ and $(x + 3) - (x + 3)(x - 7) = B(x + 3)$, then B is

   (a) $(6 - x)$ (b) $(x - 7)$ (c) $(x - 6)$ (d) $(x - 4)$ (e) $(8 - x)$

7. The graph of the function $y = f(x)$ is shown. Which of the following is true?

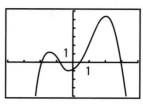

(a) $f(0) > 0$ (b) $f(2) > f(1)$ (c) $f(-1) < f(-2)$

(d) $f(-2.5) > f(0.5)$ (e) $f(-2) > f(2)$

8. In the figure shown below, the perimeter of the square is 8. What is area of triangle *ABC*?

(a) 16   (b) 2   (c) 1   (d) 4   (e) 32

9. Which of the following could be the graph of $y = \log x$?

(a) 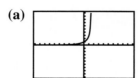   (b) 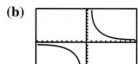   (c)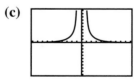

(d) 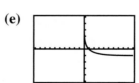   (e)

10. Solve: $-3\sqrt{2x} = 36$

(a) 72   (b) −6   (c) −72   (d) 6   (e) no real solution

Lesson 2

Polynomials and Factoring

A polynomial function is a power function or the sum of two or more power functions, each of which has a nonnegative integer power. Because polynomial functions are built from power functions, their domains are all real numbers. The graph of a polynomial function is smooth

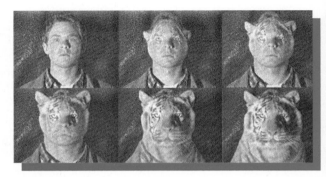

Polynomials are used to morph one image into another on Scholastic Production's Animorphs.

and unbroken. This important property of polynomial functions makes them particularly useful for transforming one image into another during special effects in movies, for creating fonts for computers, and for approximating solutions to otherwise uncomputable problems.

In this lesson, you will explore some of the advantages of expressing polynomials in different equivalent forms.

Think About This Situation

Polynomials can be written symbolically in several different forms. Consider the three forms of one polynomial shown below.

Standard Form: $P(x) = 5x^4 + 10x^3 - 65x^2 - 70x + 120$

Factored Form: $P(x) = 5(x - 1)(x + 2)(x - 3)(x + 4)$

Nested Multiplication Form: $P(x) = (((5x + 10)x - 65)x - 70)x + 120$

a What mathematical questions could be most easily answered using:
  - the standard form?
  - the factored form?
  - the nested form?

b Can every polynomial be written in factored form? In nested multiplication form?

INVESTIGATION 1 ▸ Need for Speed

In the previous lesson, you may have noticed that the time needed to produce a curve on a calculator depends upon the complexity of the equation involved. This fact is very important to computer scientists, mathematicians, and engineers, for whom computation time is often a critical component of any project. Saving computer time saves money. Super computers such as the SGI Origin 2000 at the National Center for Super-computing Applications at the University of Illinois at Champaign–Urbana can cost between 2 and 3 million dollars a year to operate.

Kray super computer

1. The form of a function rule can greatly influence the time it takes to calculate function values or produce its graph. In this activity, you will use your calculator to measure computation time for polynomial functions.

   a. Working in pairs, measure and record the time required to graph the function

   $$y = x^5 - 15x^4 + 85x^3 - 225x^2 + 274x - 120$$

   with the viewing window set to $0 \leq x \leq 6$ and $-10 \leq y \leq 10$. One person should watch the graph and the other should watch the time. Be sure to turn off any other functions or plots before doing this experiment.

   b. Next measure and record the time it takes to graph the function

   $$y = (x - 1)(x - 2)(x - 3)(x - 4)(x - 5).$$

   Be sure to turn off any other functions.

   c. Verify that the function rules in Parts a and b are equivalent.

   d. Which form produces the graph more quickly? What percentage of time is saved using this form?

   e. Computation time is measured in units called *cycles*. Suppose computing any power, such as $(1.2)^4$, uses twice as much computation time as does any multiplication, subtraction, or addition, which each require 1 cycle.

   - How many cycles are required when

     $$y = x^5 - 15x^4 + 85x^3 - 225x^2 + 274x - 120$$

     is evaluated for a particular value of x?

   - How many cycles are required when the equivalent rule

     $$y = (x - 1)(x - 2)(x - 3)(x - 4)(x - 5)$$

     is evaluated?

   - How does this computation time for the factored form compare to the computation time for the standard polynomial form?

2. The factored form of a polynomial can reduce the computation time, but not all polynomials can be written as a product of linear factors. Another way a polynomial can be rewritten to avoid exponents is to use *nested multiplication*.

   a. Verify that the polynomial $5x^5 + 4x^4 + 3x^3 + 2x^2 + x + 1$ can be rewritten in nested form as $((((5x + 4)x + 3)x + 2)x + 1)x + 1$.

   b. Determine the percentage of computation time (in seconds) saved in graphing the function $y = 5x^5 + 4x^4 + 3x^3 + 2x^2 + x + 1$ expressed in nested multiplication form rather than in standard polynomial form. Use the window $-2 \leq x \leq 2$ and $-40 \leq y \leq 40$.

   c. How many cycles are required to compute each form of the polynomial? How is the cycle count related to your answer in Part b?

   d. Investigate other viewing windows for the graph of the polynomial function. Explain why this polynomial cannot be written as a product of linear factors. Into how many linear factors can this polynomial be factored?

   e. Rewrite the polynomial of Activity 1 using nested multiplication. How does the computation time for nested multiplication compare to the computation time for the factored form?

Checkpoint

The form of a polynomial influences computation speed as well as the information that can be determined by examining the form.

ⓐ Write the polynomial function $y = (x + 2)(x - 5)(x^2 + 1)$ in standard polynomial form; in nested multiplication form.

ⓑ What information about the graph of a polynomial function can you get from the factored form? From the expanded or standard form? From the nested multiplication form?

Be prepared to discuss the importance of each form of a polynomial with your classmates.

Evaluating polynomials using nested multiplication is frequently called **Horner's Method** after the English mathematician William George Horner (1786–1837). However, evidence was published in 1911 that Paolo Ruffini (1765–1822) had used the method 15 years before Horner. More recently, the method has been found over 500 years before either Ruffini or Horner in the works of Chinese mathematicians during the late Sung Dynasty: Li Chih (1192–1279), Chu Shih-chieh (1270–1330), Ch'in Chiu-shao (c. 1202–1261) and Yang Hui (c. 1261–1275).

On Your Own

Because evaluation of polynomial functions only requires addition, subtraction, and multiplication, they are often used to estimate nonpolynomial functions. Graph the function $y = \sin x$ in the window $-4 \leq x \leq 4$ and $-2 \leq y \leq 2$. Then graph each of the following polynomial functions in the same viewing window.

i. $y = x$

ii. $y = x - \frac{x^3}{3!}$

iii. $y = x - \frac{x^3}{3!} + \frac{x^5}{5!}$

iv. $y = x - \frac{x^3}{3!} + \frac{x^5}{5!} - \frac{x^7}{7!}$

a. Compare the graphs of the polynomial functions to the graph of the sine function.

b. Determine the zeroes of each polynomial function. How do these values compare to the zeroes of the sine function?

c. Write the next polynomial function approximation in this sequence.

d. Compare the time it takes your calculator to graph the fourth polynomial function above written in standard form and in nested multiplication form.

INVESTIGATION 2 Strategic Factors

Throughout your work with quadratic and higher degree polynomial functions, the factored form of the polynomials has played an important role. Historically, writing polynomials in factored form was important for finding solutions to equations. Problems involving the solution of polynomial equations date back to the time of the ancient Egyptians, Babylonians, and Greeks. The first systematic presentation of the solution of polynomial equations is attributed by some historians to a group of Arab mathematicians starting with al-Kwarizmi (c. 800–847).

1. In her article "The Art of Algebra" (*History of Science*, June 1988, pp. 129–164), Karen Parshall describes the evolution of solving polynomial equations.

Karen Parshall

   a. She reports that a systematic solution of the equation $x^2 + 10x = 39$ was described by al-Kwarizmi in the early 800s.

    ■ Describe all the methods you know for solving this equation.

    ■ Solve this equation by reasoning with the symbolic form itself.

   b. Several centuries earlier around 250 A.D., Diophantus of Alexandria solved the equation $630x^2 + 73x = 6$. Using a method of your choice, find the solutions to this equation.

2. Until the sixteenth century, the solution of polynomial equations involved systematic guessing-and-testing for possible linear factors involving integers.

   a. Expand the polynomial $P(x) = (x - a)(x - b)(x - c)$, where a, b, and c are given constants. Write the polynomial in standard form where the coefficients of each power of x are expressions involving a, b, and c.

   b. What do you observe about the constant term? How could the constant term be used to identify possible factors of a polynomial?

   c. What do you observe about the coefficient of the x^2 term?

   d. How might you generalize your observations in Parts b and c so that they apply to nth-degree polynomials? Compare your generalization with those of other groups. Resolve any differences.

   e. Use your generalization to solve the following equations.
      i. $x^3 - 10x^2 + 27x - 18 = 0$
      ii. $x^3 - 28x = 48$
      iii. $x^3 - 4x = 0$

3. Rewrite each polynomial function in factored form using information from its graph and your generalization in Activity 2.

   a. $p(x) = x^3 + 9x^2 + 11x - 21$

   b. $y = x^4 + 2x^3 - 13x^2 - 14x + 24$

   c. $A = w^4 - 3w^3 + 2w^2$

   d. $s = t^4 + 2t^3 - 11t^2 - 12t + 36$

While searching for factors of a given polynomial, you probably used a fundamental relationship between the factors and the zeroes of the polynomial function.

Factor Theorem

The linear expression $(x - b)$ is a factor of a polynomial function $f(x)$ if and only if $f(b) = 0$.

You will be asked to complete a proof of this theorem in Organizing Task 1 (page 400).

4. The method of Activity 2 can be generalized to find factors of the form $(ax - b)$.

   a. Without expanding completely, mentally determine the leading coefficient and constant term of the polynomial $f(x) = (-1x + 1)(3x - 2)(-5x - 4)$.

   b. What are the zeroes of the polynomial function in Part a? How are the zeroes related to divisors of the leading coefficient and of the constant term of the polynomial?

c. Use your observations from Parts a and b to factor each of the following polynomials:

   i. $4x^3 + 9x^2 - 4x - 9$

   ii. $3x^3 + 11x^2 + 12x + 4$

   iii. $x^3 - 2x^2 - 4x + 8$

Your work in Activities 2 and 4 suggests the following useful theorem about possible rational zeroes of a polynomial function with integer coefficients.

Rational Zeroes Theorem

For a polynomial function with integer coefficients, if a is a divisor of the coefficient of the term of highest degree and b is a divisor of the constant term, then $\frac{b}{a}$ is a possible zero of the function, with a corresponding possible factor of the form $(ax - b)$.

---

5. You now have several ways to determine if a polynomial of the form $(x - a)$ is a factor of a polynomial $p(x)$. In this activity, you will examine a method for finding the polynomial $q(x)$ such that $p(x) = (x - a)q(x)$.

   a. Describe two ways of verifying that $(x + 1)$ is a factor of $2x^3 + x^2 - 4x - 3$.

   b. To find the polynomial which when multiplied by $(x + 1)$ gives $2x^3 + x^2 - 4x - 3$, you can use a procedure similar to long division. **Polynomial division** is illustrated below. Explain how the term in bold below is chosen at each step.

$$
\begin{array}{r}
2x^2 - x - 3 \\
x + 1 \overline{\smash{)}2x^3 + x^2 - 4x - 3} \\
\underline{2x^3 + 2x^2} \qquad\qquad \text{Multiply } (x + 1) \text{ by } \mathbf{2x^2} \\
-x^2 - 4x - 3 \qquad \text{Subtract} \\
\underline{-x^2 - x} \qquad\quad \text{Multiply } (x + 1) \text{ by } \mathbf{-x} \\
-3x - 3 \qquad \text{Subtract} \\
\underline{-3x - 3} \qquad \text{Multiply } (x + 1) \text{ by } \mathbf{-3} \\
0 \qquad \text{Subtract}
\end{array}
$$

Therefore, $2x^3 + x^2 - 4x - 3 = (x + 1)(2x^2 - x - 3)$.

   c. Verify that $2x^3 + x^2 - 4x - 3 = (x + 1)(2x^2 - x - 3)$.

   d. Show that $2x^3 + x^2 - 4x - 3$ can be written as a product of three linear factors.

   e. The polynomial $x^3 - 5x - 12$ is equivalent to $x^3 - 0x^2 - 5x - 12$.

   ■ Use polynomial division to show that $x^3 - 5x - 12 = (x - 3)(x^2 + 3x + 4)$.

   ■ Can $x^3 - 5x - 12$ be written as a product of three linear factors? Explain your reasoning.

6. Factor each polynomial into polynomials of the smallest possible degrees.

    a. $f(x) = x^3 - 2x^2 - 4x + 8$

    b. $g(x) = x^4 - 3x^3 + 2x^2$

    c. $h(x) = 2x^3 - 5x^2 - 28x + 15$

    d. $p(x) = x^3 + 3x^2 + 4x + 2$

    e. $q(x) = 8x^4 + 52x^3 + 66x^2 + 31x + 5$

7. In Lesson 1, you discovered that the multiplicity of a zero of a polynomial function was related to the shape of its graph near the zero.

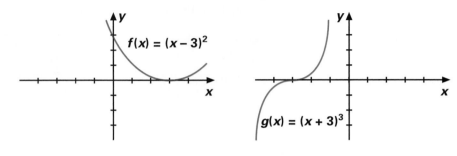

If the graph of a polynomial function touches, but does not intersect, the x-axis at $x = a$, then the zero a is repeated an even number of times. If the graph crosses the x-axis at $x = a$ and has a flattened appearance at this x-intercept, then the zero a is repeated an odd number of times.

    a. Compare the graphs of the polynomial functions in Activity 6 with their factorizations. Is the expected relationship between the shape of the graph, multiple zeroes, and repeated factors confirmed?

    b. Find a polynomial function whose graph matches the graph shown below.

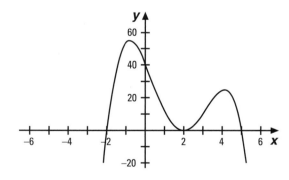

    c. Use the connection between graphs, multiple zeroes, and repeated factors to help factor each of these polynomials into polynomials of the smallest degrees possible.

       i. $y = x^3 - 7x^2 - 5x + 75$

     ii. $y = x^5 + 4x^4 + x^3 - 10x^2 - 4x + 8$

   iii. $y = x^5 + x^4 - 2x^3 - 5x^2 - 5x - 2$

    iv. $y = x^4 - 4x^3 + 13x^2 - 36x + 36$

Writing and interpreting polynomials in factored form has a long history as a central idea in algebraic thinking.

ⓐ How is the solution of a polynomial equation $p(x) = 0$ related to the factored form of the polynomial $p(x)$?

ⓑ What strategies can you use to find the factored form of a polynomial?

ⓒ How does a repeated linear factor reveal itself in the graph of a polynomial function?

ⓓ Explain why a polynomial of degree 7 cannot have just four linear factors (counting the repeated factors).

Be prepared to explain your strategies for, and thinking about, factoring polynomials to the class.

The quadratic formula can be used to solve second-degree polynomial equations. More complicated formulas exist for solving third- and fourth-degree polynomial equations. In the early 1800s, Paolo Ruffini, Neils Abel, and Evariste Galois found ways to show that it is not possible to solve fifth-degree equations by a formula. Since no general formula exists for finding roots of polynomial equations, mathematicians have developed systematic methods for estimating roots. One of these methods is investigated in the MORE set.

Neils Abel

On Your Own

As you complete these tasks, think about the relationships among zeroes, factors, and graphs of polynomial functions.

a. Factor each polynomial into polynomials of the smallest degrees possible.

- $f(x) = x^3 + 4x^2 - 11x + 6$
- $g(x) = x^3 - 3x^2 - 10x$
- $h(x) = x^4 + 5x^3 + 7x^2 + 5x + 6$

b. Sketch the graph of a fifth-degree polynomial function that has one factor repeated twice and a second factor repeated three times. Write a symbolic rule that matches your graph.

c. Write an equation for the graph shown. The y-intercept is -6.

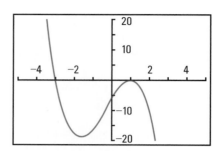

d. Andrea used the calculator graph shown at the right to help her solve the equation $x^4 - x^3 - 27x^2 + 81x - 54 = 0$. She determined that the roots of the equation were $x = 1$ and $x = 3$. Yolanda, who was working in a group with Andrea, looked at the graph and said there had to be another root and found that -6 was also a solution.

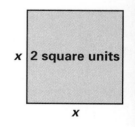

- Verify that 1, 3, and -6 are all roots.

- How did Yolanda know there had to be another root?

INVESTIGATION 3 A Complex Solution

When working on algebraic problems, it is often necessary to solve one or more equations. For thousands of years, mathematicians have struggled to create ideas and techniques for solving equations related to practical and theoretical questions. In several key cases, the solutions have required revolutionary changes in our ideas about numbers.

Pythagoras

For example, over 2,500 years ago, philosophers and scientists were puzzled by the problem of finding a square with an area of 2 square units, because the only numbers they used were positive integers and positive *rational numbers*—ratios of positive integers. Pythagoras stunned the intellectual community by proving that no rational number would provide a solution to simple quadratic equations like $x^2 = 2$. Eventually the number system was extended to include *irrational numbers* such as $\sqrt{2}$. **Irrational numbers** are real numbers that cannot be expressed as the ratio of two integers.

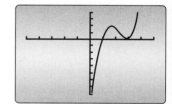

The need to extend the real numbers was prompted by the desire to solve problems like the following, which appears in the 1545 book *Ars Magna* by Cardano.

Divide 10 into two parts whose product is 40.

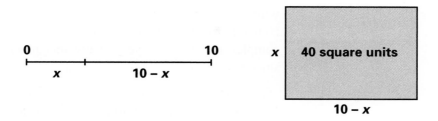

To solve this problem, Cardano needed to solve the equation $x(10 - x) = 40$, which is equivalent to $x^2 - 10x + 40 = 0$.

1. Using at least three different equation-solving strategies, search for the roots of $x^2 - 10x + 40 = 0$. Then record and interpret your findings in each case.

The solution of $x^2 - 10x + 40 = 0$ requires $\sqrt{-15}$ to be a number. But -15 does not have a square root that is a real number. To overcome this barrier, the **imaginary number** *i* was invented and defined as follows:

$$i = \sqrt{-1} \text{ or } i^2 = -1$$

Other imaginary numbers can be written as real-number multiples of *i* by using the following definition:

For any positive number b, $\sqrt{-b} = \sqrt{b}i$.

2. The solutions of the quadratic equation $x^2 - 10x + 40 = 0$ are

$$x = 5 + \frac{\sqrt{-60}}{2} \text{ and } x = 5 - \frac{\sqrt{-60}}{2}$$

Justify that these solutions can be expressed as:

$$x = 5 + \sqrt{15}i \text{ and } x = 5 - \sqrt{15}i$$

3. Numbers of the form bi, where b is a real number, are called **imaginary numbers**. Combining real and imaginary numbers, the solutions to any quadratic equation can be written in the **standard form** $a + bi$, where a and b are real numbers. Use what you know about the quadratic formula and properties of radicals to solve the following equations. Write the results in standard form.

   a. $x^2 + 6x + 34 = 0$

   b. $x^2 + 10 = 0$

   c. $2x^2 - 28x + 106 = 0$

   d. $3x^2 + 7x - 2 = 0$

   e. $x^2 - 6x + 9 = 0$

4. Examine the results from Activity 3 which involve nonreal number solutions. What do you observe about the relationship among the roots for each polynomial equation?

By the middle of the 18th century in Europe, **complex numbers** in the form $a + bi$, with a and b real numbers and $i = \sqrt{-1}$, were accepted as a new number system. Note that for any real number a, $a = a + 0 = a + 0i$. Thus, every real number is a complex number. Similarly, every imaginary number bi is a complex number since $bi = 0 + bi$.

5. An important theorem in algebra is that a polynomial of degree n has exactly n roots, counting multiplicity, within the set of complex numbers. Find the complex number roots of the following equations.

a. $x^3 - 8 = 0$

b. $x^4 - 16 = 0$

c. $2x^3 - 5x^2 + 6x = 0$

d. $3x^3 - x^2 + 4x - 6 = 0$

In the next few activities, you will explore operations on complex numbers.

6. First consider addition and subtraction of complex numbers.

a. What complex numbers in the form $a + bi$ do you think would represent the following sums and differences?

- $(4 + 6i) + (3 + 2i)$
- $(4 + 6i) - (3 + 2i)$
- $(4.5 + 7.2i) + (9 + 3.1i)$
- $(4.5 + 7.2i) - (9 + 3.1i)$
- $(12 + 9i) + (-3 - 9i)$
- $(12 + 9i) - (-3 - 9i)$

b. What general rules would show how to find the *sum* and *difference* of two complex numbers?

$$(a + bi) + (c + di) = \underline{\hspace{1cm}} + \underline{\hspace{1cm}} i$$
$$(a + bi) - (c + di) = \underline{\hspace{1cm}} + \underline{\hspace{1cm}} i$$

Compare your rules to those of other groups. Resolve any differences. Then restate in words your agreed-upon rules for adding and subtracting complex numbers.

c. Recall that for real numbers, 0 is the **additive identity** since for any real number r, $r + 0 = 0 + r = r$. What specific complex number $a + bi$ will serve as an identity element for addition of complex numbers? Prove that your answer is correct.

d. The **additive inverse** of a real number r is $-r$ since $r + (-r) = 0$. What complex number will serve as the additive inverse of the complex number $a + bi$? Prove that your answer is correct.

7. Next, consider multiplication of complex numbers.

   a. Since it makes sense to add two complex numbers by adding their *real parts* and their *imaginary parts*, you might think a similar approach would work for multiplication of complex numbers. If you use such reasoning and the fact that $i^2 = -1$ to find the product $(2 + 3i)(3 + 2i)$, what would be the result? Does your answer make sense? Explain your thinking.

   b. Another approach to multiplying complex numbers is to think about it in terms of multiplication of binomial expressions. Using this approach, what complex numbers in the form $a + bi$ would represent the products:

   - $(4 + 3i)(2 + 5i)$ ■ $(4.5 + 7.2i)(9 + 3.1i)$
   - $(3 + 2i)(4 + 6i)$ ■ $(5 + 2i)(5 - 2i)$

   If possible, check your answers with a calculator that can compute with complex numbers.

   c. What general rule would show how to find the *product* of two complex numbers?

   $$(a + bi)(c + di) = \underline{\hspace{2cm}} + \underline{\hspace{2cm}} i$$

   Compare your rule to that of other groups. Resolve any differences.

   d. Recall that for real numbers, 1 is the **multiplicative identity** since for any real number r, $1 \cdot r = r \cdot 1 = r$. What specific complex number $c + di$ will serve as an identity element for multiplication of complex numbers? Prove that your answer is correct.

   e. In some special cases, as in the fourth item in Part b, the product of two complex numbers is a real number. The complex number $5 + (-2i)$ or $5 - 2i$ is called the **conjugate** of $5 + 2i$. Prove or disprove that the product of any complex number and its conjugate is a real number.

8. Finally, consider multiplicative inverses and division of complex numbers.

   a. The **multiplicative inverse** of a nonzero real number r is $\frac{1}{r}$ or r^{-1} since $r \cdot \frac{1}{r} = \frac{1}{r} \cdot r = 1$. Use your result in Parts d and e of Activity 7 to help find the multiplicative inverse of $5 + 2i$.

   b. Find the multiplicative inverse of $\sqrt{2} + 7i$. Express your answer in standard form.

   c. Show that for any nonzero complex number $z = a + bi$, the complex number $z^{-1} = \frac{a}{a^2 + b^2} - \frac{b}{a^2 + b^2}i$ is the multiplicative inverse of z.

   d. The *quotient* of two complex numbers $z = a + bi$ and $w = c + di$ (c and d not both 0) can be defined as $\frac{z}{w} = zw^{-1}$. Express each quotient in the form $a + bi$.

   - $\dfrac{8 + i}{4 + 3i}$ ■ $\dfrac{2 + 3i}{4 - 2i}$

e. Use the definition of division in Part d and the result from Part c to derive a rule for calculating the quotient of two complex numbers $a + bi$ and $c + di$ in terms of a, b, c, and d. Compare your rule to that of other groups and resolve any differences.

9. In Activities 6–8, you established that complex numbers share some of the properties of real numbers with respect to addition and multiplication. Listed below are other real number properties with which you are familiar.

Algebraic Properties

Closure Property

of Addition:	The sum of any two real numbers is a real number.
of Multiplication:	The product of any two real numbers is a real number.

Associative Property

of Addition:	For any numbers a, b, and c: $a + (b + c) = (a + b) + c$.
of Multiplication:	For any numbers a, b, and c: $a \cdot (b \cdot c) = (a \cdot b) \cdot c$.

Commutative Property

of Addition:	For any numbers a and b: $a + b = b + a$.
of Multiplication:	For any numbers a and b: $a \cdot b = b \cdot a$.

Distributive Property

of Multiplication over Addition:	For any numbers a, b, and c: $a \cdot (b + c) = a \cdot b + a \cdot c$.
of Multiplication over Subtraction:	For any numbers a, b, and c: $a \cdot (b - c) = a \cdot b - a \cdot c$.

---

a. State corresponding properties for the complex number system and explain why each property holds.

b. One property that holds for real numbers that does *not* hold for complex numbers is the ordering property of real numbers: For any two real numbers, a and b, either $a < b$, $a = b$, or $a > b$. Consider the complex numbers 0 and i. What contradictions would arise if you assumed any one of the relationships $0 < i$, $0 = i$, or $0 > i$?

The search for solutions to polynomial equations led to the invention of complex numbers and complex number arithmetic.

ⓐ How many complex number roots will a polynomial equation of degree n have?

ⓑ For what values of a, b, and c will the equation $ax^2 + bx + c = 0$ have two distinct nonreal roots? Two distinct real roots? One real root? How will these various cases be illustrated by a graph of the function $f(x) = ax^2 + bx + c$?

ⓒ Explain why a polynomial equation of degree 3 has either zero or two nonreal complex number roots.

ⓓ Show that if $s + ti$ is a complex number zero of $g(x) = ax^2 + bx + c$, then its conjugate is also a zero of $g(x)$.

ⓔ How are addition and multiplication of complex numbers similar to, and different from, addition and multiplication of binomial expressions?

Be prepared to explain your ideas on reasoning and calculating with complex numbers.

On Your Own

The polynomial function $f(x) = x^4 - 2x^3 + 3x^2 - 2x + 2$ has complex zeroes, two of which are $1 + i$ and i.

a. Verify that $1 + i$ and i are zeroes of the polynomial function.

- Why does it follow that $1 - i$ and $-i$ are zeroes of the function?
- Write the polynomial in factored form.

b. Using the factored form of the polynomial, compute the value of the polynomial at $x = 0$, 1, and 1.5. Explain why the results must be real numbers even though the computations involve complex numbers.

Modeling

1. Describing curves solely by control points is fundamental to modern computer fonts. The mock computer font letter *A* shown below was created using parametric equations involving polynomials fitted to the (t, x) and (t, y) points in the table. The actual computer font script letter $\mathscr{A}$ was created in a similar fashion. For the actual script letter, the region between two curves was filled to create a more solid letter.

t	x	y
1	−2.8	0.3
2	1.0	4.0
3	1.4	3.5
4	0.0	0.2
5	−0.25	1.2
6	−1.8	1.8
7	1.8	1.2

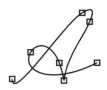

Mock Computer Font

Actual Computer Font

   a. How many horizontal changes in direction (moving from left to right or right to left) occur in tracing out this letter?

   b. How many vertical changes in direction occur in drawing this letter?

   c. What is the maximum number of vertical direction changes that can occur using two points? Using three points? Using four points? Using *n* points?

   d. What is the minimum number of points that would be necessary to draw the letter $\mathscr{A}$ shown?

   e. Verify that the graphs of the two polynomial equations $x(t)$ and $y(t)$ pass approximately through the (t, x) and (t, y) points respectively given in the table.

   $x(t) = 17.55 - 57.96419t + 58.80444t^2 - 26.56563t^3 + 6.01632t^4 - 0.67021t^5 + 0.029236t^6$

   $y(t) = 47.4 - 120.6066t + 112.50944t^2 - 48.42916t^3 + 10.49444t^4 - 1.11416t^5 + 0.04611t^6$

   f. Verify that the parametric equations in Part e produce the letter $\mathscr{A}$ shown. You may need to adjust the viewing window on your calculator.

g. Using this set of alphabetic characters, what is the minimum number of control points that would be required to make each letter of your initials?

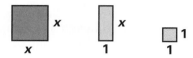

2. In Investigation 2, you were introduced to polynomial division as an aid to finding factors of a polynomial. In this task, you will investigate polynomial division using an area model. In Course 3, you used a large square to represent x^2, a long strip to represent x, and a small square to represent 1.

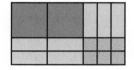

a. Verify that the following diagram represents the polynomial $2x^2 + 7x + 6$. Then rearrange the pieces to form a rectangle.

b. Explain how the following diagram illustrates the indicated polynomial division.

$$(2x^2 + 7x + 6) \div (x + 2) = 2x + 3$$

c. Shown is an area model for the polynomial $6x^2 + 13x + 6$. Explain how this model shows that when $6x^2 + 13x + 6$ is divided by $3x + 2$, the quotient is $2x + 3$ and there is no remainder.

$$(6x^2 + 13x + 6) \div (3x + 2) = 2x + 3$$

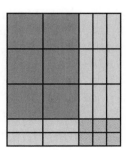

d. Create a diagram to help find the quotient:

$$(12x^2 + 17x + 6) \div (3x + 2)$$

e. As you know, division of integers sometimes results in a remainder. For example, $25 \div 4 = 6$ with a remainder of 1, since $25 = 4 \cdot 6 + 1$. Division of polynomials will also sometimes result in a remainder. Use an area model to divide $12x^2 + 17x + 8$ by $3x + 2$. What is the remainder? Verify your results using polynomial division as developed in Investigation 2 (page 387).

f. Use an area model to find $(12x^2 + 24x + 6) \div (2x + 3)$. Which term of the quotient was the easiest to determine?

3. Although 300 years ago mathematicians rejected the idea of square roots of negative numbers, today complex numbers are used extensively in many fields, including electrical engineering.

a. The *impedance* (opposition to current) Z in an electrical circuit is a combination of *resistance R* and *reactance X*. These quantities are measured in ohms and are related by the equation $Z = R + Xi$.

- If the resistance of an electrical circuit is 8 ohms and the reactance is 5 ohms, describe the impedance.

- The total impedance of a circuit is $65 + 30i$ ohms and is determined by two sources of impedance. If one source adds $40 + 19i$ ohms to the total, find the impedance of the other source.

b. The voltage E, current I, and resistance R in an electrical circuit are related by **Ohm's law**: $E = IR$. These quantities are measured in volts, amperes, and ohms, respectively.

- Find the voltage in an electrical circuit with current $2 - 5i$ amperes and resistance $3 + 4i$ ohms.

- Find the current in an electrical system with voltage $12 - 8i$ volts and resistance $7 + 5i$ ohms.

- Find the resistance in a circuit with current $6 + 4i$ amperes and voltage $8 - 11i$ volts.

4. As mathematicians searched for models to make sense of complex numbers, several proposals focused on the fact that every complex number is defined by an ordered pair of real numbers: $a + bi \leftrightarrow (a, b)$. That suggests representing complex numbers as points in a coordinate plane, called the **complex number plane**. Sometimes it is useful to represent complex numbers as vectors in standard position, as in the diagram at the right.

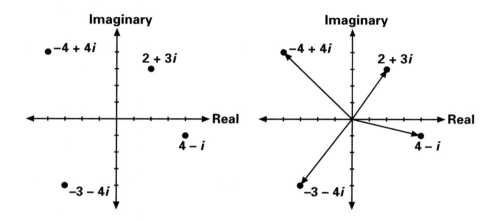

a. Graph each of the following complex numbers in the complex number plane.

   ■ $1 + 5i$ ■ $-3 + 2i$ ■ 6
   ■ $1 - 3i$ ■ $4i$ ■ $-5 - 3i$

b. How is the vector representing the sum of two complex numbers, $z + w$, related to the vectors representing $z = a + bi$ and $w = c + di$? Use graph paper to make some sketches of specific combinations of complex numbers to test and illustrate your ideas.

c. How is the vector representing the difference of two complex numbers $z - w$ related to the vectors representing $z = a + bi$ and $w = c + di$? Make some specific calculations and sketches to test and illustrate your ideas.

d. The **absolute value** of a complex number $|a + bi|$ is defined to be the length of the corresponding vector (a, b). What formula can be used in calculating $|a + bi|$? Illustrate your ideas by calculating the absolute value of the complex numbers in Part a.

e. Describe geometrically the effect of multiplying a nonzero complex number by a real number $r > 0$; by a real number $r < 0$.

f. Describe geometrically the effect of multiplying a nonzero complex number by i; by $-i$.

Organizing

1. The Factor Theorem in Investigation 2 (page 386) states a fundamental relationship between the zeroes and linear factors of a polynomial function.

   a. Explain each step in the following proof of the statement: If $(x - b)$ is a factor of $f(x)$, then $f(b) = 0$.

      i. If $(x - b)$ is a factor of $f(x)$, then $f(x) = (x - b)g(x)$, where $g(x)$ is a polynomial.

      ii. If $f(x) = (x - b)g(x)$, then $f(b) = 0$.

   b. Explain each step in the following proof of the statement: If $f(b) = 0$, then $(x - b)$ is a factor of $f(x)$.

      i. Dividing $f(x)$ by $(x - b)$ will give the equation $f(x) = (x - b)q(x) + R$, where $q(x)$ is a polynomial and R is a constant.

      ii. If $f(b) = 0$, then R must equal 0.

      iii. If $R = 0$, then $(x - b)$ is a factor of $f(x)$.

   c. Explain why both Parts a and b are needed to prove the Factor Theorem.

2. Solve the following equations, giving complex number roots in standard $a + bi$ form. Confirm the reasonableness of your solutions graphically.

   a. $x^2 + x + 11 = 0$ **b.** $x^3 + 1 = 0$

   c. $x^3 + 2x^2 + 11x + 22 = 0$ **d.** $6x^3 + x^2 - 31x + 10 = 0$

   e. $x^4 - 16 = 0$

3. The complex number system was constructed in stages that began with the set **N** of natural numbers $\{1, 2, 3, 4, \ldots\}$ and gradually introduced other important sets of numbers. Make a Venn diagram that illustrates the relationship among the following sets of numbers: natural numbers, integers, rational numbers, irrational numbers, real numbers, imaginary numbers, and complex numbers.

4. Calculate these sums and differences of complex numbers. Express answers in standard form.

   a. $(3 + 2i) + (5 + 8i)$ **b.** $(4 - 3i) + (2 + 6i)$

   c. $(8 + 10i) - (3 + 2i)$ **d.** $(-3 + i) - (7 - i)$

   e. $(5 + 2i) + (-5 - 2i)$ **f.** $(\sqrt{56} + 2i) - (4 - \sqrt{17}i)$

5. Calculate these products and quotients of complex numbers. Express answers in standard form.

   a. $5i(3 + 2i)$ **b.** $(2 - 7i)(6)$

   c. $(3 + 2i)(5 + 8i)$ **d.** $(-3 + 2i)^2$

   e. $\dfrac{4 + 9i}{3}$ **f.** $\dfrac{-3 + i}{7 - i}$

   g. $\dfrac{3 + 2i}{i}$ **h.** $\dfrac{-3 + i}{4i}$

6. Recall that for a complex number $a + bi$, the number a is called the *real part* and b is called the *imaginary part*. Two complex numbers are equal if and only if their real and imaginary parts are equal. Use this fact and properties of the complex number system to complete the following tasks.

  a. Solve $z + (3 + 2i) = 7 - 9i$ for z.

  b. Solve $z - (4 - 5i) = 12 + 3i$ for z.

  c. Show that any equation of the form $z + (a + bi) = c + di$ has exactly one solution and give the formula for that solution.

  d. Determine the two square roots of $8 - 6i$. (*Hint:* Solve $(a + bi)^2 = 8 - 6i$.)

Reflecting

1. Look back at the proof of the Factor Theorem outlined in Organizing Task 1. Explain why if a polynomial $p(x)$ is divided by $x - a$, then the remainder is $p(a)$. This result is often referred to as the **Remainder Theorem**.

2. The speed benefit of nested multiplication is not widely known, even though it works for any polynomial. On the other hand, most mathematics students are familiar with factoring polynomials. Why do you think factored form is more heavily stressed than nested multiplication form?

3. Consider all quadratic equations of the form $ax^2 + bx + c = 0$.

  a. For what combinations of positive and negative signs of a and c can you be sure that the equation will have only real number roots?

  b. How can you tell from the quadratic formula that such an equation will have either two nonreal complex number roots or two real number roots— never just one complex number root?

4. In the historical development of the number systems, several ideas (in addition to square roots of negative numbers) proved difficult barriers to progress. As late as the 16th and 17th centuries, irrational numbers like $\sqrt{17}$, and negative numbers like -5 were challenged as meaningless, absurd, false, and impossible. Why do you suppose that acceptance of such numbers took so long?

5. Think back to when you first encountered negative numbers or irrational numbers. Can you remember being puzzled about any aspects of those new mathematical ideas, saying to yourself, "This doesn't make any sense to me"? What does it take for things in mathematics to "make sense" to you?

6. Powers of i: i^2, i^3, i^4, i^5, i^6, i^7, ... have an interesting periodic property. Using the fact that $i^2 = -1$, rewrite each power of i in simplest form and look for a pattern that would allow you to quickly rewrite powers like i^{100} or i^{523}.

Extending

1. In solving a new mathematical problem, it is often helpful to reduce it to a simpler problem by temporarily ignoring some details. Consider the task of solving a fourth-degree (quartic) equation like $x^4 + 6x^2 - 16 = 0$.

   a. Rewrite the given equation, replacing x^2 by u. What variable expression should replace x^4?

   b. Solve the new equation for u.

   c. Now use the relationship $x^2 = u$ to solve for x.

   d. Use this method to solve the following equations:

   - $2x^4 + 3x^2 - 2 = 0$
   - $x^6 + x^4 - 4x^2 - 4 = 0$

   e. What features of the three equations in x above provide clues to the usefulness of a *substitution of variable*?

2. In the "On Your Own" of Investigation 1, you investigated the following four polynomial function approximations to the function $y = \sin x$.

   i. $y = x$ **ii.** $y = x - \frac{x^3}{3!}$

   iii. $y = x - \frac{x^3}{3!} + \frac{x^5}{5!}$ **iv.** $y = x - \frac{x^3}{3!} + \frac{x^5}{5!} - \frac{x^7}{7!}$

   a. Graph the *error function* (the difference between the polynomial function and the sine function) for each polynomial in the window $-4 \leq x \leq 4$ and $-0.1 \leq y \leq 0.1$. For example, the first error function is $y = x - \sin x$. What does the shape of the graph of the error function tell you about the degree of the next polynomial approximation and the sign of its leading coefficient?

   b. What patterns do you see in this sequence of error functions?

3. Extending Task 2 of Lesson 1 introduced synthetic substitution as an efficient paper-and-pencil method of evaluating a polynomial function at a particular number. The Remainder Theorem (Reflecting Task 1, page 401) suggests a relationship between polynomial division and polynomial evaluation. In this task, you will explore a useful connection between the processes of synthetic substitution and polynomial division. Consider the polynomial division and the synthetic substitution shown below.

$$
\begin{array}{r}
2x^2 - x - 3 \\
x + 1 \overline{)2x^3 + x^2 - 4x - 3} \\
\underline{2x^3 + 2x^2} \\
-x^2 - 4x - 3 \\
\underline{-x^2 - x} \\
-3x - 3 \\
\underline{-3x - 3} \\
0
\end{array}
$$

$$
\begin{array}{r|rrrr}
-1 & 2 & 1 & -4 & -3 \\
   &   & -2 & 1 & 3 \\
\hline
   & 2 & -1 & -3 & 0
\end{array}
$$

a. How is the fact that -1 is a zero of $p(x) = 2x^3 + x^2 - 4x - 3$ indicated in each of these examples?

b. How do the coefficients of the quotient polynomial and the remainder in polynomial division show up in the synthetic substitution process?

c. Test your ideas by finding the quotient and remainder for $(3x^3 - 7x - 4) \div (x + 1)$.

d. Use synthetic substitution to rewrite each of the following expressions.

    i. $(2x^3 - 5x^2 - 3x + 4) \div (x + 1)$

    ii. $\dfrac{2x^3 - 6x + 4}{x + 2}$

    iii. $(6x^3 + 5x^2 + 6x - 5) \div (2x - 1)$

e. How could you use synthetic substitution to find linear factors of a polynomial?

f. Why do you think synthetic substitution is sometimes called "synthetic division"?

4. One method for estimating zeroes of a polynomial function is called the **bisection algorithm**. The method uses the simple observation that if you know two points $(a, f(a))$ and $(b, f(b))$, with $f(a)$ and $f(b)$ having different signs, then a zero must lie between a and b.

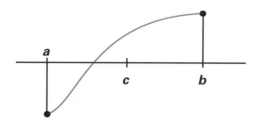

To find the zero, calculate the midpoint c between a and b. Either $f(c) = 0$ or the sign of $f(c)$ matches the sign of either $f(a)$ or $f(b)$. If $f(c) = 0$, then the zero has been found and the algorithm is completed. Otherwise, if $f(a)$ and $f(c)$ have the same sign, then let $a = c$, else let $b = c$. Repeat the process for the new values of a and b.

a. Explain why the above algorithm locates a zero.

b. Using the bisection algorithm, write a calculator program to estimate the zeroes (to four decimal places) of a polynomial function.

c. Use your program to estimate the zeroes of $y = x^3 - 2.4x^2 + 2$.

d. How could the initial values of a and b be determined by the program?

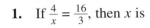

1. If $\frac{4}{x} = \frac{16}{3}$, then x is

   (a) $\frac{3}{4}$　　　　(b) $\frac{3}{64}$　　　(c) $\frac{20}{3}$　　　(d) $\frac{7}{16}$　　　(e) $\frac{4}{3}$

2. Simplify: $\frac{2}{R} - \frac{7}{T}$

   (a) $\frac{-5}{R-T}$　　　(b) $\frac{-5}{RT}$　　　(c) $\frac{2T-7R}{RT}$　　　(d) $\frac{5}{R+T}$　　　(e) $\frac{2R-7T}{RT}$

3. The graphs of the two equations $2x + 6y = 7$ and $6x + 2y = 9$ are

   (a) two perpendicular lines　　(b) two distinct parallel lines
   (c) the same line　　　　　　(d) two intersecting lines that are not
   (e) not straight lines　　　　　　　perpendicular

4. If $f(x) = \frac{7}{x-2}$, for what value of x does $f(x) = 5$?

   (a) 7　　　　(b) 3　　　　(c) 3.4　　　　(d) 2　　　　(e) 1.4

5. Which of the following is the equation of a quadratic function whose graph has x-intercepts of 11 and -2?

   (a) $f(x) = x^2 - 13x - 22$　　　(b) $f(x) = x^2 + 9x - 22$

   (c) $f(x) = x^2 - 9x - 9$　　　(d) $f(x) = (x-11)^2 - 2$

   (e) $f(x) = x^2 - 9x - 22$

6. Solve: $\frac{3}{8}(2x - 5) = \frac{3}{8}$

   (a) $x = 2.5$　　(b) $x = -2.5$　　(c) $x = -2$　　(d) $x = 3$　　(e) $x = 2$

7. Which of the following could be a portion of the graph of $y = -\frac{2}{x}$?

(a) 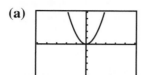　　**(b)** 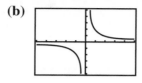　　**(c)**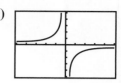

(d) 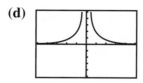　　**(e)**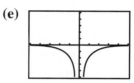

8. In the figure shown, $\sin \theta =$

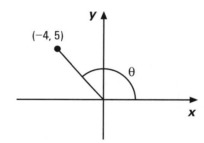

(a) $-\frac{5}{4}$　　**(b)** $\frac{\sqrt{41}}{5}$　　**(c)** $-\frac{4}{5}$　　**(d)** $\frac{-4}{\sqrt{41}}\pi$　　**(e)** $\frac{5}{\sqrt{41}}$

9. If $2^x = 22$, then which of the following best approximates x?

(a) $x = 4.46$　　**(b)** $x = 11$　　**(c)** $x = 4.64$　　**(d)** $x = 1.04$　　**(e)** $x = 1.64$

10. Solve: $3\sqrt[a]{x} = 5$

(a) $x = 2^a$　　**(b)** $x = \frac{5}{3} - a$　　**(c)** $x = \left(\frac{5}{3}\right)^a$　　**(d)** $x = 8^a$　　**(e)** $x = \left(\frac{3}{5}\right)^a$

Rational Functions

Birds such as sea gulls and crows feed on various types of mollusks. To break the shell of a mollusk, they lift it into the air and drop it onto rocks below. Biologists have observed that northwestern crows drop a type of mollusk called a *whelk* from a fairly consistent height of about 5 meters. The crows are rather selective in that they appear only to pick up large-sized whelks. They are also persistent—one crow was observed dropping a whelk 20 times before it finally broke. Scientists have suggested that this behavior is an example of optimization in nature.

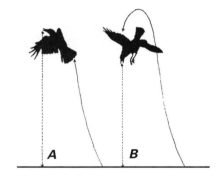

Possible Flight Paths

Why do northwestern crows drop whelks from a height of about 5 meters?

Think About This Situation

Think about the process crows might use to break open a whelk.

ⓐ Which flight path, *A* or *B*, do you think the birds use? Why?

ⓑ What factors do you think might influence the height at which the birds choose to drop a whelk?

ⓒ Do you think there is a minimum or maximum number of drops required to break a whelk? A minimum or maximum height at which a whelk can be dropped to break?

ⓓ Sketch what you believe to be the graph of the number of drops required to break a whelk as a function of the height of the drop. How are your answers to Part c evident in your graph?

INVESTIGATION 1 Building Models

In this investigation, you will explore whether crows drop whelks at a height that minimizes the amount of *work* to break them open. You might expect that the amount of work to break open a whelk depends upon the number of times a whelk has to be dropped and the height of the drop. The number of drops also likely depends on the height of the drop. So a step in finding a mathe-

matical model for the work involved is to search for a relationship between the height of a drop and the number of drops required to break open a whelk.

1. To investigate this situation, the Canadian biologist Reto Zach dropped a whelk from a fixed height until it broke. He recorded the number of drops needed to break open a whelk from that height. He then repeated this proce-dure for several whelks and calculated the average number of drops for a given height. The results for several different heights are given in the fol-lowing table. Make and examine a scatterplot of these data.

Height of Drop (m)	2	3	4	5	6	7	8	10	15
Average Number of Drops	55	10	7.5	6.0	5.0	4.3	3.6	3.0	2.5

Source: Zach, Reto. Selection and dropping of whelks by northwestern crows. *Behaviour*, Vol 67 (1978): pp. 134–147.

   a. Is there a minimum or a maximum number of drops required to break open a whelk? Justify your answers.

   b. Is there a minimum height at which a whelk will break? Explain your thinking.

   c. What does the shape of the graph suggest about a basic function model relating the two variables?

   d. At this point in the investigation, what heights would you recommend be used for breaking open a whelk by dropping? Explain your reasoning.

Note that the number of drops required to break open a whelk does not vary greatly as the height of the drop gets large. When a function $y = f(x)$ tends to have values closer and closer to a line as x or y gets infinitely large (or small), the function is said to be **asymptotic** to that line, and the line is called an **asymptote**. You have seen asymptotes before in the context of inverse power functions $y = \frac{a}{x^b}$, where b is a positive real number. For example, the function $y = \frac{1}{x}$ has the x-axis ($y = 0$) as a horizontal asymptote and the y-axis ($x = 0$) as a vertical asymptote.

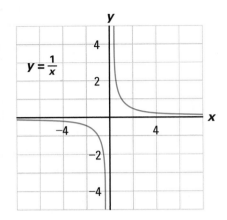

2. Compare the graph of the (*height, average number of drops*) data from the whelk-dropping experiment to the graph of the function $y = \frac{1}{x}$.

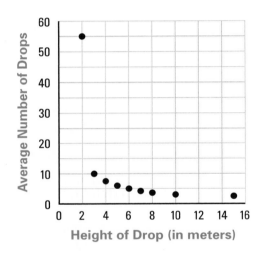

a. What is a possible horizontal asymptote for the plot of the whelk data?

b. How is the existence of a horizontal asymptote related to the physical situation? Start by completing the sentence, "As the height of the drop gets large, the average number of drops... ."

c. Do you think there is a vertical asymptote for data arising from the whelk-breaking experiment? Why or why not?

d. Based on your previous work with translation of function graphs, explain how the graph of $N(x) = a + \frac{c}{x - b}$ is related to the graph of $y = \frac{1}{x}$.

e. These data can be modeled by a function of the form $N(x) = a + \frac{c}{x - b}$.

  ■ What value do you think is appropriate for a? Explain your answer.

  ■ What do you think is a reasonable range of values for b? Explain your reasoning.

f. In your previous work, you often used the regression capabilities of your graphing calculator or computer software to fit functions to data. Which regression models can you immediately rule out for these data? Explain why each of the remaining regression models would not be appropriate to fit a function to these data in their current form. (A systematic method for finding appropriate values of b and c through linearizing the data is developed in Extending Task 1.)

The function $N(x) = a + \frac{c}{x - b}$ relating average number of drops to height of drop is an example of a *rational function*. A **rational function** can be expressed as the ratio or quotient of two polynomial functions. For example, the function $f(x) = 3 + \frac{1}{x^2}$ is a rational function since it can be rewritten as $f(x) = \frac{3x^2 + 1}{x^2}$.

3. Verify that the function $N(x) = a + \frac{c}{x - b}$ modeling the (*height, average number of drops*) data can be rewritten as $N(x) = \frac{ax + k}{x - b}$ for some constant k.

4. When objects are lifted, two important factors influence the amount of work being done. They are the weight of the object and the distance the object must be lifted. Think about the work done by a crow in dropping whelks.

a. Suppose a crow lifts one whelk 3 meters, a second whelk 6 meters, and a third whelk 9 meters. Which lift required the most work if all three whelks were the same size? Which required the least? About how much more work do you think was required?

b. Suppose one crow dropped 10 whelks from 3 meters, while another crow dropped 5 whelks of the same size from 6 meters. Which crow do you think did more work? Why?

c. The northwestern crows only drop large whelks which are about ten times the weight of small whelks. How much more work is done by crows dropping large whelks instead of small whelks from 4.5 meters?

d. What relationship do you think exists between the amount of work W done, the weight F of the whelks, the distance D the whelks are lifted, and the number of lifts N?

The work done by the northwestern crows will depend upon the weight of the whelk, the height of the drop, and the number of drops required to break open the whelk. In the case of the crows, however, the weight of the whelk is fairly uniform since they only feed on the largest whelk. Thus, the amount of work will be a function of the height of the drop and the number of drops.

One reasonable conjecture is that the crows select a height that minimizes the work they have to do to break open a whelk. Another conjecture is that they select a height that minimizes the loss to other crows waiting on the ground to steal their food while also trying to get as high as possible to ensure that the whelk breaks. By modeling the behavior mathematically, you can investigate why the crows consistently fly up to about 5 meters.

5. One function model which fits the data for the average number of drops required to break a whelk at various heights is $N(x) = 1 + \frac{20}{x - 0.94}$.

   a. What does this function rule indicate about the minimum number of drops required to break open a whelk? The minimum height at which a whelk will break?

   b. For which data points in the table on page 407 does this model fit least well? How might the design of the experiment explain your observation?

   c. Express the work done in breaking open a whelk as a function of drop height x. Sketch a graph of your work function W.

   d. Is there a best height from which to drop a whelk in order for it to break? If so, what is it?

   e. What does the result suggest about why northwestern crows consistently drop large whelks from a height of about 5 meters?

   f. What do you observe about the amount of work needed to break open a whelk for very large heights?

   g. Rewrite your function rule for the work done in dropping a whelk so that the rule clearly represents a rational function. Indicate what two polynomials are involved.

6. You now have expressed the work function in two forms. One form makes the method of calculating work as the product of the number of drops and height of the drop explicit. The second form represents the expression for work as a rational function. By rewriting the function rule for work in different forms, you can get a better understanding of the relationship between the work done and the height from which the whelk is dropped.

   a. A third form of the work function rule can be found by division. Verify that $W(x) = \frac{x^2 + 19.06x}{x - 0.94}$ can be rewritten as $W(x) = (x + 20) + \frac{18.8}{x - 0.94}$.

   b. Compare the graph of the work function to the graph of the function $f(x) = x + 20$ for large values of x. Interpret the similarities based on the situation.

   c. What do you observe about the amount of work for heights just above the minimum height?

   d. Graph your function for the work together with the function $y = \frac{18.8}{x - 0.94}$. For values of x near the minimum height, how do the two functions compare?

   e. Show that the work function $W(x)$ is the sum of the two functions $y = x + 20$ and $y = \frac{18.8}{x - 0.94}$.

f. Explain why the portion of the graph for positive heights to the left of the vertical asymptote should not be considered.

g. Summarize how rewriting the work function as a sum of two functions was useful in more completely interpreting the work required as a function of height.

In the case of the whelk experiment, the function $y = x + 20$ represents the general behavior of the work function for the majority of values of x. In this case, the line $y = x + 20$ is called an **oblique asymptote** that is, an asymptote that is neither horizontal nor vertical. This fact indicates that as the height increases, and thus the likelihood that only one drop is necessary, the amount of work is just the work done in lifting the whelk to the chosen height. Near the vertical asymptote, the rational function $y = \frac{18.8}{x - 0.94}$ has the greatest influence over the behavior of the work function $W(x)$. This fact indicates that if you choose a height which requires multiple drops, the work is essentially determined by the number of times you have to drop it.

7. In this investigation, you have seen that rational functions can be written in several different forms. Each of these forms reveals important information about the function. For each of the following functions:

  i. Rewrite the function in a form that clearly indicates that the function is a rational function.

  ii. Rewrite the function in a form that allows you to determine the oblique asymptote of the function.

  a. $f(x) = x + \frac{5x}{x - 1}$

  b. $g(x) = x\left(1 + \frac{2}{x - 3}\right)$

Checkpoint

Look back at your work in modeling the data from the whelk-dropping experiment.

ⓐ What features of a situation suggest a rational function model rather than a polynomial or exponential function model?

ⓑ How can thinking about a problem context suggest the asymptotes which might arise when modeling the situation?

ⓒ At least three different ways of writing a rational function for the work involved in dropping a whelk were used in this investigation. What information does each form provide?

ⓓ Give examples of rational functions, different from those in the investigation, whose graphs have horizontal, vertical, or oblique asymptotes. Be sure to indicate the asymptote(s) in each case.

Be prepared to discuss your ideas and examples with the entire class.

On Your Own

Optimal packaging is important to every industry. Many factors including ease of handling, public appeal, ease of construction, and durability influence the design of containers. Food items such as sodas, puddings, soups, fruits, and vegetables are packaged in cans that are right circular cylinders. Often, the amount a can is to hold has been established by tradition or by consumer desires. Consider a soft drink can that holds 355 milliliters (ml) or 355 cubic centimeters (cc) (or 12 fluid ounces) of drink.

a. Assuming a can must hold 355 ml of liquid and the general shape is a right circular cylinder, what relationship must exist between the radius r and the height h of the container?

b. One major cost in manufacturing a can is the amount of material needed to make it. Determine an appropriate formula for the surface area of a right circular cylinder.

c. Using your relationship from Part a, write a rule expressing surface area of the can as a function of its radius.

d. Explain why there must be an optimal radius and height to minimize the surface area of the can by estimating the surface area for very large and very small values of the radius.

- Find the radius which produces the smallest surface area.
- Determine the corresponding height.

e. Look back at your function modeling the surface area of a can with volume 355 ml.

- Explain why the function is a rational function.
- What are the asymptotes of the graph of this function?
- What is the significance of each asymptote in modeling the shape of a can?

INVESTIGATION 2 ▶ Rationalizing It All

Patterns in data plots and problem conditions as they relate to asymptotic behavior often suggest rational functions as appropriate models. In this investigation, you will explore rational functions and their asymptotes more generally.

1. Carefully consider the function $f(x) = \frac{2(x + 1)}{(x + 1)(x - 1)}$ and its graph.

    a. What are the domain and range of this function?

b. Graph the function using the **ZDecimal** window on your calculator. Examine the table of values and graph for this function.

- How is the domain of the function reflected in the table? In the graph?

- How is the range of the function reflected in the table? In the graph?

c. Write equations in the form "$x = ...$" for any vertical asymptotes of this function. How are the vertical asymptotes related to the equation for the function?

d. Write equations in the form "$y = ...$" for any horizontal asymptotes of this function. How are the horizontal asymptotes related to the equation for the function?

e. In Lesson 1, you examined the end behavior of polynomial functions. How is the end behavior of rational functions related to asymptotes?

2. Working together in pairs, carefully consider each of the following functions. Record a graph of each function, along with your work.

$$\textbf{i.} \ \ y = \frac{2}{x+1} \qquad \textbf{ii.} \ \ y = \frac{x+2}{x+5} \qquad \textbf{iii.} \ \ y = \frac{3x}{x^2-9}$$

$$\textbf{iv.} \ \ y = \frac{x^2-25}{x-3} \qquad \textbf{v.} \ \ y = \frac{2x^2+1}{x^2+1} \qquad \textbf{vi.} \ \ y = \frac{x^3-1}{x^2-1}$$

$$\textbf{vii.} \ \ y = \frac{x^3+x+1}{x+3} \qquad \textbf{viii.} \ \ y = \frac{x^2+2x-3}{x-1}$$

a. For each function:

- State the domain and range.

- Find the zeroes of the function.

- Determine the end behavior of the function.

b. Next look for asymptotes.

- Write equations for any horizontal asymptotes.

- Write equations for any vertical asymptotes.

- Write equations for any oblique asymptotes.

c. Use the graphs of the functions to find:

- Any local maxima or minima

- Any absolute maximum or minimum

3. As a group, compare and examine your results in Activity 2.

a. How can you find the zeroes of a rational function by examining its symbolic form as a quotient of polynomials?

b. How can you predict the horizontal, vertical, or oblique asymptotes of a rational function by reasoning with its symbolic form?

c. What characteristics of a rational function can you use to help you predict the behavior of the function near a vertical asymptote?

d. How can you predict the end behavior of a rational function?

4. Using the ideas of zeroes, y-intercept, asymptotes, and end behavior, explain why the graph below matches the behavior of the function $f(x) = \dfrac{15(x+3)^2}{(x-1)(x-5)}$.

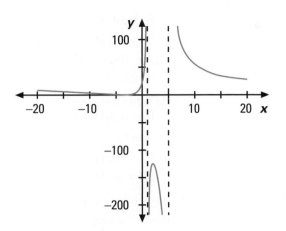

5. Determine a possible equation for each of the following graphs by examining the behavior of the graph near zeroes and asymptotes.

a.

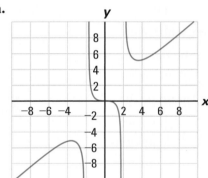

b.

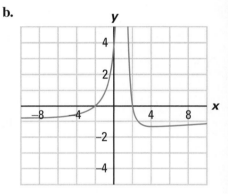

6. Each of the following functions has at least one asymptote. Indicate what you think are the asymptotes for each function and explain why each function is not a rational function.

a. $y = \tan x = \dfrac{\sin x}{\cos x}$

b. $y = \dfrac{\cos x}{x}$

c. $y = \dfrac{\sin x}{x}$

d. $y = \dfrac{2^x}{x+3}$

e. $y = \dfrac{x}{|x|}$

In this investigation, you discovered important connections between the graph and the symbolic form of a rational function.

a Suppose $f(x) = \frac{p(x)}{q(x)}$ is a rational function.
- What can you conclude if $p(a) = 0$ and $q(a) \neq 0$?
- What can you conclude if $q(a) = 0$ and $p(a) \neq 0$?
- What can you conclude if $p(a) = 0$ and $q(a) = 0$?
- Describe the end behavior of $f(x)$.

b How can you use the information in Part a to help sketch the graph of $y = f(x)$? What characteristics determine the behavior of the graph of the function near its asymptotes?

c How can you determine the equation for any vertical asymptote? For any horizontal asymptote? For any oblique asymptote?

d Why must you be careful when you are viewing and interpreting a calculator- or computer-produced graph of a rational function?

Be prepared to explain your group's conclusions about rational functions and their graphs.

Investigation of particular rational functions has been an important part of mathematical history, and some of these functions have been given special names, as identified in the following "On Your Own." The curve in Part a is attributed to Isaac Newton (1642–1727) who is well known for his early work on calculus. However, his teacher Isaac Barrow (1630–1677) has not received much recognition. Barrow was the first to determine the general solution for the equation of the tangent line for the kappa curve in Part b of the "On Your Own" on the next page. The Italian mathematician Maria Gaetana Agnesi (1718–1799) investigated the general form of

Maria Gaetana Agnesi

the function $y = \frac{4a^3 - ax^2}{x^2 + 4a^2}$. In her early twenties, she wrote the first text that included both differential and integral calculus. By age 30, she was an honorary member of the faculty at the University of Bologna. The curve in Part c of the "On Your Own" arose naturally as part of a geometric investigation of circles and has come to be called a *versiera*.

For each function, use symbolic reasoning to determine the zeroes of the function. Determine any vertical, horizontal, or oblique asymptotes. Check your answers using a graphing calculator or computer.

a. Newton's Serpentine: $y = \dfrac{4x}{x^2 + 1}$

b. Barrow's kappa: $y = -\dfrac{x^4}{x^2 - 1}$

c. *Versiera* of Agnesi: $y = \dfrac{4 - x^2}{x^2 + 4}$

MORE
Modeling • Organizing • Reflecting • Extending

Modeling

1. An airplane trip from Chicago to Philadelphia reported a flight time of 1 hour, 48 minutes covering 678 miles. The return trip required 2 hours 16 minutes. A major factor accounting for the difference in time was the direction of the wind. A similar phenomenon is also evident to swimmers and boaters in a flowing river. Suppose your swimming speed in still water is *v*.

   a. The Beaver Creek river flows at approximately 1.5 meters per second.

   ■ What is your velocity swimming downstream with the flow of the river?

   ■ What is your velocity on the way back swimming against the flow?

   ■ Write a rule for predicting the total time *t* required to swim 100 meters down river and back.

   b. Write a rational function rule for the total time required to travel a distance d downstream and back in a river flowing at a rate v_r. Examine your function rule.

     ■ What does it indicate if your velocity is equal to the speed of the river? Explain what this means in terms of your trip.

     ■ What happens if your velocity is less than the speed of the river?

     ■ What is the total travel time if your velocity is 0? Explain why this value does not make sense and what is wrong.

   c. Determine the speed of the plane and the wind speed for the Chicago-Philadelphia trip described at the beginning of this task.

2. In your previous work, you saw that the weight of an object above Earth can be modeled by a function of the form $w(h) = \left(\frac{r}{r+h}\right)^2 w_0$, where r is the radius of Earth (approximately 3,950 miles), h is the distance above Earth's surface, and w_0 is the weight of the object on the surface of Earth.

   a. Using this model, explain why the force exerted by the crows in lifting a whelk can be considered constant over the likely heights at which the birds fly.

   b. At what height will a "150-pound object" weigh 50 pounds?

   c. How high would a crow carrying a whelk have to fly for the whelk's weight to change by 0.1%?

3. Rational functions are used extensively in the field of optics. In the diagram below, if F is the focal length, in millimeters, of a lens, u is the distance between the object and the lens, and v is the distance between the lens and the film (image), then $\frac{1}{F} = \frac{1}{u} + \frac{1}{v}$. For this task, assume that the lens has a focal length of 50 millimeters.

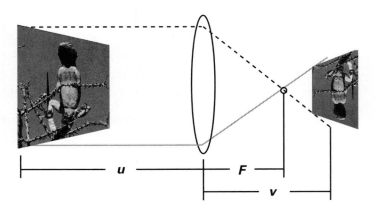

a. Rewrite the equation $\frac{1}{F} = \frac{1}{u} + \frac{1}{v}$ to express v as a function of F and u.

b. If the camera is focused on an object 6 meters away, what is the distance between the lens and the film? What if the object is 0.5 meters away?

c. For an object at a distance u units away, the distance v between the lens and the film is adjusted so that the object is in focus. Objects at other distances which produce similar values for v will also be in focus. The range of distances for which objects will be in focus is called the "depth of field." Examine the graph of v as a function of u. Describe the behavior of the graph in terms of the distance between the object and the lens and the distance between the film and the lens.

d. Explain why the depth of field is large when you focus on objects that are very far from the lens, but the depth of field is small for objects close to the lens.

4. Cereal boxes are typically made from a single piece of cardboard with specified lines for cuts and folds. One appropriate design is shown on the next page with dotted lines representing fold lines and solid lines representing cuts. In this model, a 0.5-inch tab is used to glue the box together.

For cereal boxes, a good design is one that wastes as little cardboard as possible while maximizing the volume. Suppose a company wishes to calculate the optimal design of a box under these conditions. Assume that 720 square inches of cardboard are available.

 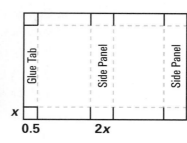 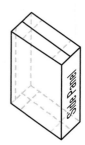

a. Explain why the width of the side panels has to be twice the width of the flaps.

b. Label the other unknown dimensions of the box and write equations for the surface area and volume of the box.

c. For cereal boxes, the width of the box should not be too large. Assume the box should have a width of 2 inches for easy handling. Taking into account this fact and the fixed surface area, what two equations must the height and length satisfy?

d. What is the layout that produces a maximum volume? Explain why you know you have found the optimal design.

e. In cereal boxes, the width of the side panels is slightly less than twice the width of the flaps so that the top flaps can be tucked in and the bottom flap glued. How can you adjust your solution method to take this added restriction into account?

5. Although cereals vary in shape and weight, most cereal boxes have similar design. Boxes that must hold heavy objects have other structural constraints. To reinforce the bottom and overall structure of a box, a shipping company specifies that the length of the front of a box is to be three times the width of the flaps. The side panels must be twice the width of the flaps and include a 0.5-inch tab for gluing. A limit of 1,400 square inches has been placed on the layout.

a. By sketching the "folded" box, explain why you think this design is better for boxes holding heavy objects than the cereal box design in Modeling Task 4.

b. Find modeling equations for the area and volume of this box.

c. Use your modeling equation for volume to find the dimensions of the box with maximum volume. Justify why you know that you have optimized the volume.

Organizing

1. Look back at your work for the "On Your Own" (page 412) of Investigation 1.

   a. Record the optimal height and radius of the can in Part d of that task.

   b. Construct mathematical models to minimize the surface area of cans with volumes of 236 ml, 650 ml, and 946 ml.

   c. Plot the optimal height against the optimal radius for each of the four volumes.

   d. What relationship seems to exist between the optimal radius and height for a given volume?

2. Determine a possible equation for each of the following graphs by examining the behavior of the graph near zeroes and asymptotes.

   a.
   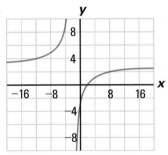

   b.
   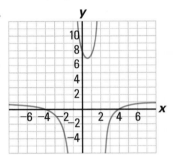

   c.
   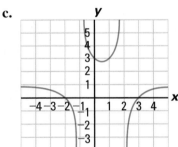

   d.
   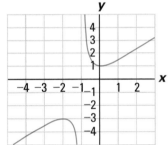

3. Rational functions arise naturally when working with similar triangles. Investigate the following problem of determining the maximum length of an object that will fit horizontally around a corner with 8-foot wide hallways. Keep in mind that this length is actually the smallest of all possible lengths L that satisfy the conditions of the diagram at the right.

   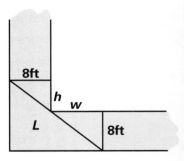

   a. To avoid square roots that often arise when modeling distance problems, it is common practice to maximize the square of the distance $S = L^2$. Explain why doing so will produce the same solution as maximizing the distance L.

   b. Using similar triangles, determine a relationship between h and w.

c. Write a function rule relating the distance L^2 to the lengths h and w. Then use the relationship from Part b to write L^2 as a function of h.

d. Determine the minimum value of L^2 and L.

e. What angle is made by the object with one of the walls? Explain how this angle is determined by the symmetry of the situation.

f. What is the minimum value of L if one of the hallways is only 4 feet wide?

4. The average velocity calculations introduced in Unit 1, "Rates of Change," often produce expressions that look like rational functions.

a. The height above the water of this Mexican cliff diver is modeled well by the function $h(t) = 30 - 4.9t^2$.

■ Write an equation for the average velocity between time $t = a$ and time $t = b$.

■ Interpret the meaning of the equation for average velocity in terms of the graph of $h(t)$ and the two points $(a, h(a))$ and $(b, h(b))$.

b. Consider finding the average velocity between the point 1 second into the dive and the point b seconds into the dive.

■ Explain why the average velocity is a rational function expressed in the variable b.

■ Explain why the average velocity exists for any value of b other than $b = 1$.

■ Examine the graph of the average velocity function for values of b near 1. As b gets close to 1, the average velocity gets close to approximating the instantaneous velocity. Use this to explain why the rational functions that arise from calculating average velocity often are undefined at a point, but do not have an asymptote at that point.

5. Without graphing, identify all zeroes and asymptotes for each function. Then, make a sketch of the graph of the function. Check your graph by comparing it to one produced by a graphing calculator or computer.

a. $f(x) = \frac{2x^2 - 4x - 6}{x^2 + x - 6}$

b. $f(x) = \frac{x^2 - 9}{(x - 2)^2}$

c. $f(x) = \frac{2x^2 + 2}{x^2 - 1}$

d. $f(x) = \frac{x^2 - x - 12}{x - 4}$

Reflecting

1. In studying the behavior of the northwestern crow, Reto Zach dropped a number of whelk of various sizes from a range of heights. The resulting data are given in the following plots. The curves are those that Reto Zach sketched by hand.

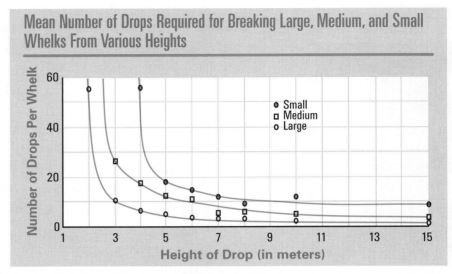

Mean Number of Drops Required for Breaking Large, Medium, and Small Whelks From Various Heights

Source: Zach, Reto. Selection and dropping of whelks by northwestern crows. *Behaviour*, Vol 67 (1978): pp. 134–147.

a. What do you observe about the vertical and horizontal asymptotes of the graphs corresponding to the different sizes of whelks?

b. What plausible physical explanations can you give for the differences in the asymptotes?

2. Think about the work involved in moving sandbags in an effort to control flooding.

a. Why do you think people form a line to transport the sandbags?

b. Suppose the people carried the bags one-by-one from the source of the bags to the area of the flooding. Would more or less work be done in moving the sandbags in this manner? Explain your answer.

3. It is common to see data patterns that have close to an L-shape as in the plot below. In such situations, you might consider modeling the data with a rational function $y = \frac{ax + b}{x - c}$.

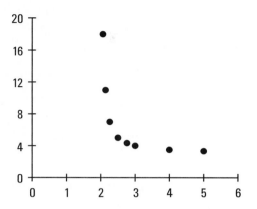

   a. How is the value of a related to the graph of the function?

   b. How is the value of b related to the graph of the function?

   c. How is the value of c related to the graph of the function?

4. For each of the following descriptions, draw a sketch of the graph of a possible function and write the function rule.

   a. The function has one zero, a vertical asymptote $x = 2$, and a horizontal asymptote $y = -1$.

   b. The function has two zeroes, vertical asymptotes $x = -1$ and $x = 2$, and horizontal asymptote $y = 3$.

   c. The function has one zero, vertical asymptotes $x = -3$ and $x = 2$, and the x-axis as the horizontal asymptote.

5. Under what conditions will the end behavior of a rational function be asymptotic to a line?

Extending

1. In Unit 3, "Logarithmic Functions and Data Models," you studied how linearizing data can be used to help determine the most appropriate regression model for a set of paired data. In Investigation 1, the conjecture was made that the data from the whelk-dropping experiment might be modeled by an equation of the form $N(x) = a + \frac{c}{x - b}$. Examine the following procedure for linearizing the data assuming this model.

   a. Why would $a = 1$ be a reasonable assumption in the case of dropping whelk?

   b. The function rule $N(x) = 1 + \frac{c}{x - b}$ can be rewritten as $\frac{1}{N(x) - 1} = \frac{1}{c}(x - b)$. Keeping in mind that b and c are constants, explain why a linear relationship should exist between x and $\frac{1}{N(x) - 1}$.

c. Using the experimental data in Investigation 1, complete a table like the one shown below.

Height of Drop x (cm)	Average Number of Drops ($N(x)$)	$\dfrac{1}{N(x)-1}$
2		
3		
4		
⋮		
15		

d. Using your calculator or computer software, make a scatterplot of the $\left(x, \dfrac{1}{N(x)-1}\right)$ data. Use linear regression to find an equation relating $\dfrac{1}{N(x)-1}$ and x.

e. Solve your equation in Part d for the number of drops $N(x)$ at height x. Then graph the equation along with the scatterplot for $(x, N(x))$. How well does the equation fit the data?

f. Where does the graph of this function have a vertical asymptote?

g. Based on the data, is there a minimum drop height below which a whelk will not break?

h. The first step to linearizing the whelk data was to assume that $a = 1$. What problems arise in the procedure for linearizing the data if no value of a is assumed?

2. Determine a possible equation for each of the following graphs by examining the behavior of the graph near zeroes and asymptotes.

a.

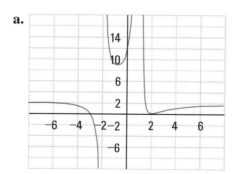

b.

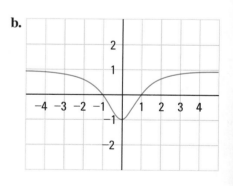

3. Consider a rational function of the form $f(x) = \frac{ax+b}{cx+d}$.

   a. What is the domain for f?

   b. Find an expression for $f^{-1}(x)$. What kind of function is f^{-1}? What is the domain of f^{-1}?

   c. Under what conditions on a, b, c, and d is the inverse function the same as the original function?

   d. Suppose $g(x) = \frac{mx+n}{px+q}$. Find an expression for $g(f(x))$. What kind of function is $g(f(x))$?

4. In Unit 1, "Rates of Change," one equation that was used to model the position of a bungee jumper was $h(t) = 32 + \frac{80 \cos (2t - 0.7)}{t+1}$. Although this function is not a rational function, it has traits similar to those of a rational function.

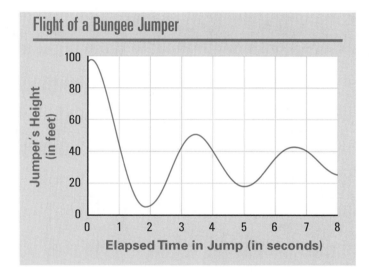

Flight of a Bungee Jumper

Jumper's Height (in feet) vs. Elapsed Time in Jump (in seconds)

   a. What are the horizontal and/or vertical asymptotes of this function? What explanation can you give for the existence of the asymptotes?

   b. How can you determine the asymptotes by reasoning with the symbolic form of the equation?

   c. A common model for *damped oscillations* is given by the equation $p(t) = 32 + 75(0.6)^t \cos (2t - 0.7)$. How does the graph of this function compare with the graph of the function used to model the position of the bungee jumper?

   d. Examine more closely the function in Part c.

   ■ Does this function have either horizontal or vertical asymptotes?

   ■ How can you determine the horizontal or vertical asymptotes by examining the equation?

1. A theater gives senior citizens a 30% discount on tickets every Thursday. If a senior's ticket costs $28.00, how much does a regular theater ticket cost?

    (a) $58.00   **(b)** $36.40   **(c)** $42.00   **(d)** $40.00   **(e)** $47.60

2. Simplify: $\left(\frac{3x-4}{2}\right) - (x+2)$

    (a) $2x$ **(b)** $\frac{x}{2}$ **(c)** $\frac{x}{2} - 8$ **(d)** $x - 4$ **(e)** $\frac{x-8}{2}$

3. For what value of b will this system have an infinite number of solutions?

    $\begin{cases} y = bx - 3 \\ 3x - y = -7 \end{cases}$

    (a) -1 **(b)** $-\frac{1}{3}$ **(c)** 3 **(d)** none   **(e)** $\frac{1}{3}$

4. The graph of $y = f(x)$ is shown to the left. Which of the following is the graph of $y = -|f(x)|$?

    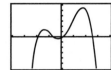

    (a) 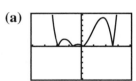   **(b)** 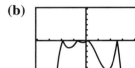   **(c)**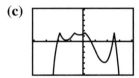

    (d) 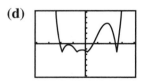   **(e)**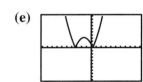

5. If the graph of a quadratic function is tangent to the x-axis, what can you determine about its zeroes?

    (a) You can determine it has two different real zeroes.
    (b) You can determine one zero is real and one is nonreal.
    (c) You can determine it has one unique zero.
    (d) You can determine it has two nonreal zeroes.
    (e) You cannot determine anything.

6. Which graph is the graph of the inequality $y \leq 2x + 5$?

(a)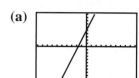

(b)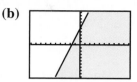

(c)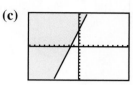

(d)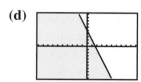

(e)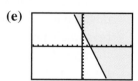

7. Which is not a root of the polynomial equation $x^4 + 4x^3 - 17x^2 - 24x + 36 = 0$?

(a) -6 **(b)** -2 **(c)** 1 **(d)** -4 **(e)** 3

8. In the figure shown at the right, $AD = 10$, $BE = 3$, and $DE = 5$. Find DC.

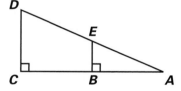

(a) 6.17   **(b)** 8   **(c)** 6

(d) 4   **(e)** 7.2

9. If $\log x = 3.5$, which of the following best approximates x?

(a) $x = 32{,}000$ **(b)** $x = 320{,}000$ **(c)** $x = 3{,}200$

(d) $x = 35$ **(e)** $x = 320$

10. If $A = P(1 + r)^t$, $A > 0$, and $P > 0$, then $r =$

(a) $\dfrac{(A - P)}{t} - 1$ **(b)** $\dfrac{\sqrt[t]{P - A}}{P}$ **(c)** $\sqrt[t]{A - P}$

(d) $\sqrt[t]{\dfrac{A}{P}} - 1$ **(e)** $\sqrt[t]{A - P} - 1$

Looking Back

In this unit, you investigated the behavior of polynomial and rational functions. By understanding the behavior of these families of functions, you are able to develop useful models for physical situations. You also examined complex numbers, their properties, their connections with polynomial functions, and some of their applications to electrical circuits. The next several tasks will help you review and synthesize important ideas in this unit and connect them to concepts and methods developed in previous units.

The first three tasks explore the optimum brood size (number of offspring) for chickadees. The central question is

How many eggs should a bird lay in order to maximize the chance of its offspring surviving?

The data reported below were collected in woods near Oxford, England by C. M. Perrins. The hypothesis is that the brood size has evolved to maximize the number of birds that survive for three months past hatching.

Only two of the several factors associated with brood size are considered here, the weight of the young at age 15 days and the percent that survive 3 months. Certainly the larger the brood size, the more chicks that might survive. However, if the brood size is too large, the adults cannot adequately feed the young and the probability of survival is reduced. Your goal is to develop a model of the number of offspring that survive as a function of the number of offspring.

1. Perrins collected data on the weight of chicks at 15 days for various brood sizes. The average weight at each brood size is shown in the table on the next page.

   a. Does the data support the conjecture above about brood size and the ability of the parent bird to adequately feed the chicks?

   b. Find a model that your group thinks best describes the relationship between brood size and weight at 15 days. Justify why you selected that particular model.

   c. For what range of brood sizes is your model appropriate?

Average Weight of Chicks of Different Brood Sizes

Brood Size	Weight in Grams at 15 Days	Brood Size	Weight in Grams at 15 Days
2	19.1	8	18.6
3	19.2	9	18.4
4	19.2	10	18.1
5	19.1	11	17.7
6	19.0	12	17.4
7	18.8	13	16.9

Source: C. M. Perrins, Population fluctuations and clutch-size in the great tit, Parus Major L., *Journal of Animal Ecology:* 34: pp. 601–647 (1965).

2. Selected chicks were then monitored until 3 months of age. The probability of survival for 3 months was examined as a function of the weight at 15 days.

Weight in Grams at 15 Days	Probability of Survival for 3 Months
17.0	0.031
17.4	0.042
17.7	0.052
18.0	0.064
18.4	0.081
18.7	0.101
19.0	0.112
19.2	0.123

a. Why do you think information on only chicks at the indicated weights was reported?

b. What does the observed data suggest about a chick's likelihood for survival?

c. Why does this table contain weights different from those reported in Task 1?

d. Find a model which you think best represents the relationship between weight at 15 days and probability of survival for three months.

e. Based on the data in the two tables, make a conjecture about the brood size that would produce the greatest number of surviving offspring.

3. In Tasks 1 and 2, you found relationships between brood size and weight at 15 days and between weight at 15 days and probability of survival for 3 months.

    a. Find a relationship between brood size and probability of survival.

    b. Using your relationship, write a rule which gives the expected number of chicks surviving to 3 months as a function of the brood size.

      ■ What is the optimal brood size so that the maximum number of chicks survive?

      ■ What is the likely number of chicks surviving at the optimal brood size? Explain the significance of this number in terms of the survival of the species.

    c. C. M. Perrins found that the average brood size is 8.1 chicks. How does the optimal brood size which you found compare to Perrins' value? Discuss the conjecture that the brood size has evolved so as to maximize the likelihood of survival.

4. Next think about the pattern of change in light intensity (the amount of light per square foot) as you walk past a lamppost.

    a. Describe how the light intensity varies when you are very far from the light and when you are right next to the lamppost.

    b. Sketch a graph of the pattern of change in your perception of the intensity of the light as you walk from a point to the left of the lamppost, to the front of the lamppost, and then to the right of the lamppost.

    c. In Course 2, you saw that the light intensity from a light source could be modeled by an equation of the form $I = \frac{k}{d^2}$ where d is the distance from the source and k is a constant. Why is this equation reasonable?

    d. Of course, the distance from the light is not just the distance from the lamppost. Choose a reasonable height for a lamppost. Then write a function rule relating the light intensity reaching the ground and the true distance from the light source. Assume $k = 100$ watts.

    e. Compare the graph of this light intensity function to your sketch.

5. Departments of transportation must decide how far apart to place lampposts to guarantee a minimum light intensity depending upon the location. To understand the light intensity with multiple lampposts, begin by considering the light intensity from two lights along a road.

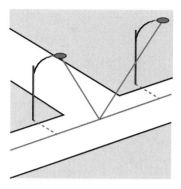

   a. Think about the pattern of change in light intensity as you approach, drive through, and leave the intersection.

   ■ Where do you think the minimum and maximum light intensities would occur?

   ■ Sketch a graph showing the pattern in change in light intensity that you would expect. If your graph includes asymptotes, indicate why you believe each asymptote must exist.

   b. Decide what would be a reasonable distance between the two lampposts. Then assuming as in Task 4 that $k = 100$ watts, write function rules for the light intensity from *each* lamp. Write a rule giving the *total* light intensity at any spot.

   ■ Graph the total light intensity function.

   ■ Is there a minimum or maximum light intensity between the lampposts? If so, where does each occur?

   ■ Explain why your function can never be zero assuming both lights are on.

   c. Show that for any choice of post heights and distance between lampposts there is no spot with zero light intensity.

6. Using two different methods, find a polynomial function whose graph passes through the points $(-1, -3)$, $(0, 0)$, $(1, -3)$, and $(2, 3)$.

7. For each of the following polynomial functions:

- Find all zeroes, including those that are complex numbers.
- Find all local maxima and minima.
- Find any absolute maximum or minimum.
- Describe its end behavior.
- Draw a sketch of its graph.

a. $f(x) = 5x^2 - 6x + 4$

b. $g(x) = x^3 + 10x^2 - 69x + 90$

c. $h(x) = 6x^4 + 11x^3 - 11x^2 - 14x + 8$

d. $i(x) = x^3 + 9x^2 + 28x + 20$

e. $j(x) = (x - 1)^3(x + 2)^2$

8. Classify each of the following statements as *True* or *False*. If true, explain your reasoning. If false, provide a counterexample.

a. Every real number is a complex number.

b. The product of two imaginary numbers is an imaginary number.

c. The sum of a complex number and its conjugate is a real number.

d. A polynomial function with real coefficients always has an even number of nonreal complex number zeroes.

e. Counting multiplicity of zeroes, a polynomial function of degree n has n complex number zeroes.

9. Perform the indicated complex number operations. Express your answers in standard form.

a. $(3 - 4i) - (7 - 2i)$

b. $(2 + i)(3 - i)$

c. $(a + bi)(a - bi)$

d. $\dfrac{1 + i}{1 - i}$

10. For each of the following rational functions:

- Find all zeroes.
- Determine all asymptotes.
- Describe its end behavior.
- Draw a sketch of its graph.

a. $y = \dfrac{2x + 1}{x - 3}$

b. $y = \dfrac{x - 4}{x^2 + x - 2}$

c. $y = \dfrac{4x^2 - 8x - 12}{x^2 + 3x + 2}$

d. $y = \dfrac{x^2 + 10}{x^2 - 2x - 8}$

e. $y = \dfrac{x^3 - x^2 + 1}{x^2 - 2x + 1}$

f. $y = \dfrac{-3x - 8}{x^2 + 5x + 6}$

11. Determine a possible equation for each of the following graphs.

a.

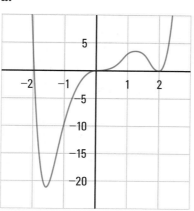

b.

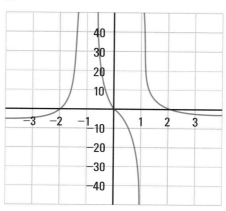

Checkpoint

In this unit, you modeled problem situations using polynomial and rational functions and developed an understanding of some of the properties of these families of functions. You also examined the complex number system.

a What characteristics of a problem situation suggest that a polynomial or a rational function might be a useful model?

b What are the general features of graphs of polynomial functions?

c How are the linear factors, zeroes, and graphs of a polynomial function related?

d What are the general features of graphs of rational functions? How are these features revealed in the different symbolic forms of rational function rules?

e What are complex numbers, and how are they related to real numbers and to imaginary numbers? How do they arise in solving polynomial equations?

f Describe how you add, subtract, multiply, and divide two complex numbers.

Be prepared to explain your responses and thinking to the entire class.

On Your Own

Write, in outline form, a summary of the important mathematical concepts and methods developed in this unit. Organize your summary so that it can be used as a quick reference in future units and courses.

Functions and Symbolic Reasoning

Reasoning with Exponential and Logarithmic Functions

In previous units and courses, you have solved problems involving quantitative variables by examining tables and graphs of relationships among the variables and by operating on the symbolic expressions for the relationships. In this unit, you will review and extend your skills in *symbolic manipulation*—rewriting expressions in equivalent forms, often to solve equations—and in *symbolic reasoning*—making inferences about symbolic relations and about relationships between symbolic representations and graphical, numerical, and contextual representations.

Think back to the data from Unit 3, "Logarithmic Functions and Data Models," relating average distance from a planet to the Sun and the number of Earth days needed for that planet to orbit the Sun.

Planet	Mercury	Venus	Earth	Mars	Jupiter	Saturn	Uranus	Neptune	Pluto
Average Distance from Sun *d* (in millions of miles)	36	67	93	142	484	887	1,783	2,794	3,661
Orbit Time *t* (in Earth days)	88	225	365	687	4,333	10,759	30,685	60,189	90,465

To make an intelligible graph of the data, it is helpful to plot the logarithms of the distance and time values (*log average distance*, *log orbit time*). The plot is a linear pattern, and the regression equation for the transformed data is $\log t = 1.50 \log d - 0.39$.

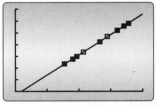

Transformed Data

INVESTIGATION 1 ▶ Exponents and Exponential Functions

The planetary data in the "Think About This Situation" can be modeled by both log $t = 1.50$ log $d - 0.39$ and $t = 0.41d^{1.5}$. Yet these representations don't look alike at first glance. In Unit 3, you learned how to use fundamental ideas of exponents and logarithms and symbolic reasoning to show the two representations are equivalent. This investigation and the three that follow will help you sharpen your algebraic skills in transforming exponential and logarithmic equations into equivalent equations so that they may be more easily interpreted and solved.

You have used exponential functions to model relationships between pairs of variables such as time and growth of colonies of bacteria, time and the value of investments, and number of rebounds and rebound height of a dropped ball. In the next several activities, you will revisit familiar exponential models with particular attention to the information conveyed by the form of their symbolic rules.

1. In earlier work, you saw that liquids cool in predictable patterns. For example, if ice is added to a glass containing a soft drink, the temperature (in degrees Celsius) of the soft drink at the bottom of the glass can be expressed as a function of time (in minutes) by the rule $f(t) = 23(0.94^t)$.

   a. What do the numbers 23 and 0.94 represent in this situation?

   b. How long will it take for the liquid at the bottom of the glass to cool to 10°C? Explain how your answer can be determined by analysis of:

   - the graph of $f(t)$;

   - a table of values for $f(t)$;

   - the symbolic rule for $f(t)$.

   c. Determine the number of *seconds* it will take before the liquid at the bottom of the glass has cooled to 5°C. Explain the reasoning you used to solve the problem.

2. A class at Jefferson High School conducted an experiment in which rubber balls were dropped onto the floor from a height of 8 feet and the rebound heights were tracked. For one of the balls, the class found that the height of each rebound was about 90% of the height from which the ball fell.

   a. Write a rule expressing the height of the ball as a function of the number of bounces.

   b. How high did the ball bounce on the twelfth bounce?

   c. How many bounces were there before the ball rebounded no higher than 4 in.?

   d. Compare your modeling equation in Part a and solution method in Part c with those of other students. Resolve any differences.

3. Recall that the general form of the equation of an exponential function is $f(x) = a(b^x)$, where $a \neq 0$ and $b > 0$, $b \neq 1$. The parameter a is the initial value (at $x = 0$) and the parameter b is the growth or decay factor.

   a. What are the values of a and b for the model you developed in Activity 2?

   b. What are the domain and range for exponential functions?

   c. Write a recursive formula that is equivalent to $f(x) = a(b^x)$ for $a \neq 0$, $b > 0$, $b \neq 1$, and x a nonnegative integer.

4. Suppose that the size of a population is modeled by $f(t) = 100\left(8^{\frac{t}{6}}\right)$, where t is measured in months.

   a. What is the size of the initial population?

   b. How can you determine that the population increases by a factor of 8 every 6 months.

   c. How long would it take for the population to double?

   d. What is the monthly rate of growth? The yearly rate of growth?

5. The form of the rule for the exponential function in Activity 4 made it easy to determine the growth rate for 6-month periods. Rewriting the rule in equivalent forms can provide additional information.

   a. Manuel provided the following symbolic reasoning as part of his justification that the population in Activity 4 doubles every two months.

   ■ Evaluate his reasoning and provide supporting reasons for each step.

   $$\text{Since } f(t) = 100\left(8^{\frac{t}{6}}\right)$$
   $$f(t) = 100\left(2^3\right)^{\frac{t}{6}}$$
   $$f(t) = 100\left(2^{\frac{3t}{6}}\right)$$
   $$f(t) = 100\left(2^{\frac{t}{2}}\right).$$

   ■ How did this form of the function rule enable Manuel to conclude that the population doubles every two months?

   b. Write $f(t) = 100\left(8^{\frac{t}{6}}\right)$ in the form $f(t) = 100(b^t)$. What monthly growth rate is revealed by this symbolic form?

   c. Write $f(t) = 100\left(8^{\frac{t}{6}}\right)$ in a form that shows the yearly growth rate.

   d. Write $f(t) = 100\left(8^{\frac{t}{6}}\right)$ in the form $f(t) = 100(10^{kt})$.

Your work in Activity 5 illustrates some of the power of symbolic reasoning and manipulation. Transformation of an equation modeling a situation into an equivalent form often provides new information about the model. For example, if the annual growth of a population is given by $g(t) = 55\left(9^{\frac{t}{4}}\right)$, then $g(t)$ can be rewritten as:

$$g(t) = 55\left(3^{\frac{t}{2}}\right) \qquad \text{(to establish that the population triples every 2 years)}$$
$$= 55\left(\sqrt{3}\right)^t \qquad \text{(to show the yearly growth rate)}$$
$$= 55\left(10^{0.2386t}\right) \qquad \text{(to show base 10 representation)}$$

6. Basic properties of exponents are useful in rewriting exponential functions in equivalent forms. In this activity, you will review these properties.

   a. Write each expression in an equivalent form with a single exponent. Where appropriate, describe any necessary restrictions on the variable in the base.

   i. $5^3 \cdot 5^7$ **ii.** $\pi^5 \div \pi^2$ **iii.** $3^{k-2} \cdot 3^{k+5}$

   iv. $(\pi^5)^2$ **v.** $x^3 \div x^7$ **vi.** $(x+1)^{3.1} \cdot (x+1)^{4.6}$

   vii. $(z-3)^{3.1} \div (z-3)^{4.6}$ **viii.** $[(p+3)^{3.1}]^{4.6}$ **ix.** $\left[(z-2)^{2k}\right]^{\frac{k-1}{2}}$

   x. $(x-2)^{p+4} \cdot (x-2)^{p-3}$

   b. Examine your work in Part a. Group Items i–x based on whether they are applications of either the *Product of Powers Property*, the *Quotient of Powers Property*, or the *Power of a Power Property*. Compare your groupings with those of your classmates and resolve any differences.

Checkpoint

In this investigation, you reviewed modeling with exponential functions and discovered advantages of rewriting exponential rules in equivalent forms.

ⓐ Often data exhibiting an exponential growth (decay) pattern are modeled by a rule of the form $f(t) = a(c^{kt})$ because the form suggests a particularly useful or easily understood interpretation. What information is evident in each of the four symbolic forms in Activity 5?

ⓑ Explain how a function rule of the form $f(t) = a(c^{kt})$, where a, c, and k are constants, can be rewritten in the form $f(t) = a(b^t)$.

ⓒ Suppose that a colony of five bacteria grows exponentially by a factor of 25 every four hours.

   ■ Write a rule for the growth of this bacteria colony as a function $g(t)$ of time in hours. What do the input and output for this function represent?

   ■ How would you transform the rule for $g(t)$ into a form from which the hourly growth rate can be read? What is that growth rate?

   ■ How would you transform the rule for $g(t)$ into a form involving a base of 5? Interpret the equation.

   ■ How would you transform the rule for $g(t)$ into a form with a base of 10?

Be prepared to compare your responses and thinking to that of other groups.

Consider the exponential function $h(t) = 10\left(64^{\frac{t}{12}}\right)$.

a. Describe a possible context that could be modeled by this function. Interpret the variables in this context.

b. Express the rule for $h(t)$ in a form involving a base of 8. What growth rate information for your identified context is provided by this form?

c. Express the rule for $h(t)$ in a form involving a base of 4. What growth rate information for your context is provided by this form?

d. If t represents months of growth for a population, what equivalent expression for $h(t)$ would reveal the growth rate for every two-month period?

e. Express $h(t)$ in the form $h(t) = a(b^t)$.

INVESTIGATION 2 Solving Exponential Equations

Since exponential growth and decay patterns are quite common, it is important to have efficient strategies for reasoning with symbolic representations of exponential models. Finding the output of an exponential function $f(x) = a(b^x)$ is straightforward. A calculator will give you a value for b^x given any nonnegative b and any x, as long as b^x is not too large for the calculator to handle. Then you need only multiply by a. The process of finding a value x such that $f(x) = c$, that is, solving $a(b^x) = c$ for x is more cumbersome. In this investigation, you will examine symbolic procedures which can always be used to solve exponential equations.

1. The number of hosts on the Internet has increased quite rapidly over the last 20 years. In fact, during this period the Internet has changed the look of education, business, and communication around the world. The number of Internet hosts as tracked by the Internet Software Consortium (www.isc.org) can be modeled reasonably well by the exponential function $f(x) = 230(2.04^x)$, where x is time in years and $x = 0$ corresponds to 1981.

:: Domain Registration

:: Domain Transfers

:: Free Site Builder

:: Web Hosting

:: Web Site Promotion

:: Online Tutorials

:: Internet Access

a. What do the numbers 230 and 2.04 reveal to you about this situation?

b. What number of Internet hosts does the model give for your year of birth? For 1995? For the year 2007? On what assumption is your last prediction based? Do you think this is reasonable?

c. On your own, devise a procedure to find the year in which the number of Internet hosts reached 1 million according to this model.

d. Compare your procedure with that used by others. Do all procedures give equally accurate results? Are all procedures equally easy to carry out?

2. When students in a class at Calvin High School compared procedures for solving $230(2.04^x) = 1,000,000$, they were surprised at the variety of methods used. Identify the properties of logarithms or equality that justify each step in Maria's solution.

$$230(2.04^x) = 1,000,000$$
$$2.04^x = \frac{1,000,000}{230}$$
$$\log 2.04^x = \log\left(\frac{1,000,000}{230}\right)$$
$$x \log 2.04 = \log 1,000,000 - \log 230$$
$$x = \frac{\log 1,000,000 - \log 230}{\log 2.04}$$
$$x \approx 11.75$$

Thus, the year is 1981 + 11.75 or 1992.75. So, according to this model, the number of Internet hosts reached 1 million in 1992.

a. Is Maria's solution correct? If so, could her method be used for similar problems? Why or why not?

b. In the second step, she took the base ten logarithm (log) of each side of the equation. Why does equality continue to hold when you take the logarithm of both sides of an equation?

c. Rebecca used a procedure very similar to Maria's, except that in the third and fourth steps she reasoned:

$$x \log 2.04 = \log\left(\frac{1,000,000}{230}\right)$$
$$x = \frac{\log\left(\frac{1,000,000}{230}\right)}{\log 2.04}$$

Do you see any advantages or disadvantages to Rebecca's method?

3. Emmet, another student in the class, offered a different way to solve $230(2.04^x) = 1,000,000$. He reasoned as follows:

$$230 \cdot 2.04^x = 1,000,000$$
$$\log (230 \cdot 2.04^x) = \log 1,000,000$$
$$\log 230 + \log 2.04^x = \log 1,000,000$$
$$x \log 2.04 = \log 1,000,000 - \log 230$$
$$x = \frac{\log 1,000,000 - \log 230}{\log 2.04}$$
$$x \approx 11.75$$

Justify each step in Emmet's solution. Is his solution correct?

4. Maria, Rebecca, and Emmet each used some important properties of logarithms to solve the exponential equation. For each of the following properties, identify where, if at all, the property was used above.

  i. The solutions of $\log g(x) = \log h(x)$ are also solutions of $g(x) = h(x)$.

  ii. $\log \frac{a}{b} = \log a - \log b$

  iii. $\log (a \cdot b) = \log a + \log b$

  iv. $\log a^m = m \log a$

5. Activities 2 and 3 illustrated the usefulness of logarithms as an equation-solving tool. It is important, however, not to lose sight of the general functional characteristics of logarithms.

   a. Sketch the graph of $f(x) = \log x$.

   b. What is the domain of $f(x) = \log x$? The range?

   c. Why does it make sense that x must be greater than 0?

6. In earlier work, you saw that investments in which interest is compounded can be modeled using exponential functions. For example, the future value of $1,000 invested at 4% interest compounded annually is given by $V(t) = 1,000(1.04^t)$, where t is time in years. One of the benchmarks for evaluating investments is the time it will take for the value to double.

   a. How many years will it take for an initial investment of $1,000 to double in value if 4% interest is compounded annually?

   b. How does doubling time change if the initial investment is $2,000? $3,000? $5,000?

   c. How does the doubling time change if the interest rate is 6%? 9%? 15%?

   d. Look back over your results in Parts a–c to formulate an answer to the question, "How do initial investment and/or interest rate affect doubling time of an investment earning compound interest?"

   e. Solve the equation $a(b^t) = 2a$ for t and explain how the result justifies your answer to Part d.

   f. The tripling time of an exponential growth process is found by solving $a(b^t) = 3a$. Show how this equation can be solved for t when $b > 1$ and explain how the result shows that tripling time is independent of initial amount.

   g. Find a formula for the value of t for which $f(t) = 10(4^t)$ is n times its initial value.

In the previous activities, you saw some of the usefulness of logarithms in solving exponential equations and, more generally, in reasoning about exponential functions. The next activity will provide you additional practice with equation-solving applications of logarithms.

7. Use symbolic reasoning to solve each equation for the variable x, if possible. If not possible, explain why it is not possible.

   a. $10^x = 5$ **b.** $8^x = 20$

   c. $2.7^x = 18$ **d.** $4.1^x = -2$

   e. $a^x = 42,\ a > 0$ **f.** $42^x = a,\ a > 0$

8. Solve the general equation $r^x = s$ for x, where $r > 0$, $r \neq 1$, and $s > 0$. Why are these restrictions on the values of r and s necessary?

9. Next, consider how transformations of the variable affect the solution of exponential equations.

  a. Compare the following two methods for solving $8^{3x} = 20$. Justify each step in each solution.

<table>
<tr><td align="center">Solution I</td><td align="center">Solution II</td></tr>
<tr><td align="center">$8^{3x} = 20$</td><td align="center">$8^{3x} = 20$</td></tr>
<tr><td align="center">$(8^3)^x = 20$</td><td align="center">Let $t = 3x$; then $8^t = 20$.</td></tr>
<tr><td align="center">$x \log 8^3 = \log 20$</td><td align="center">So $t = \frac{\log 20}{\log 8}$.</td></tr>
<tr><td align="center">$x = \frac{\log 20}{\log 8^3}$</td><td align="center">But $t = 3x$, so $3x = \frac{\log 20}{\log 8}$.</td></tr>
<tr><td align="center">$x \approx 0.48$</td><td align="center">$x = \frac{\log 20}{3 \log 8} \approx 0.48$</td></tr>
</table>

  b. Check that $x \approx 0.48$ is the solution.

  c. In Solution II, how does substitution of the variable t for $3x$ simplify the problem at hand?

  d. Solve the general equation $a^{kx} = b$ for x where $a > 0$, $a \neq 1$, and $b > 0$.

  e. Solve $7^{x+3} = 41$ for x.   **f.** Solve $7^{x-3} = 41$ for x.

  g. Illustrate two different methods for solving the general equation $a^{x+c} = b$ for x, where $a > 0$, $a \neq 1$, and $b > 0$.

Checkpoint

In this investigation, you explored methods for solving variations of the general exponential equation $a(b^x) = c$, where $a > 0$ and $b > 0$, $b \neq 1$.

ⓐ In general, how would you solve an equation of the form $a(b^x) = c$ for x using symbolic reasoning?

ⓑ Examine this incomplete solution of an exponential equation. Identify the property of logarithms or equality that justifies each step in the solution process.

$$9(8^{4x - 1.7}) = 13$$
$$8^{4x - 1.7} = \frac{13}{9}$$
$$\log 8^{4x - 1.7} = \log \left(\frac{13}{9}\right)$$
$$(4x - 1.7) \log 8 = \log \left(\frac{13}{9}\right)$$
$$4x - 1.7 = \frac{\log \left(\frac{13}{9}\right)}{\log 8}$$

  ▪ Complete the solution of this equation.

  ▪ How could you check the solution?

ⓒ If $g(t) = a(b^t)$, explain how to find the value of t for which $g(t)$ is n times an initial value at t_0.

Be prepared to explain your methods and reasoning to the entire class.

Use symbolic reasoning to solve each of the following equations.

a. $84(0.75^t) = 38$

b. $5(4^{2x}) = 22$

c. $300 = 15\left(8^{\frac{t}{3}}\right)$

d. $2(7^{2x+1}) - 3 = 8$

INVESTIGATION 3 Natural Logarithms

Recall from Unit 3 that logarithms with a base of 10 are called *common logarithms*. They are widely used because the base is the same as the base for the familiar decimal numeration system. Powers of 10 have logarithms that are integers, and numbers between consecutive powers of 10 have logarithms between consecutive integers.

Any given positive number can be written in the form b^r, where b is any positive number other than 1. Since a logarithm is simply an exponent on a positive base, other base logarithms are possible. For example, $\log_3 81 = 4$ because 4 is the exponent on 3 that gives 81, that is, $3^4 = 81$. In general, **$\log_b x = y$ if and only if $b^y = x$.**

1. Consider how you might evaluate $\log_{2.7} 8$ and, more generally, $\log_b c$ for $b > 0$, $b \neq 1$, and $c > 0$.

   a. Estimate $\log_{2.7} 8$. How can you check your estimate?

   b. Write and solve an exponential equation equivalent to $x = \log_{2.7} 8$.

   c. Evaluate $\log_{3.5} 10$, $\log_{45} 13$, and $\log_6 18$. Look for a pattern in your method.

   d. Write a formula for calculating $\log_b c$ for $b > 0$, $b \neq 1$, and $c > 0$. Compare your formula to that of other groups. Resolve any differences. Then prove that the formula always works.

   e. The formula you established in Part d is called a **change of base** formula. Why do you think the formula has that name?

   f. Use the change of base formula to compute each logarithm.

   i. $\log_4 7$ **ii.** $\log_{12.3} 5$

   iii. $\log_{0.3} 6$ **iv.** $\log_\pi 2$

Now that you know how to compute logarithms with bases other than 10, you can investigate patterns in the graphs of logarithmic functions with varying bases.

2. On the same axes, sketch graphs of $f(x) = \log x$, $g(x) = \log_2 x$, $h(x) = \log_3 x$, and $j(x) = \log_{100} x$.

  a. How are the graphs of $g(x)$, $h(x)$, and $j(x)$ related to the graph of $f(x)$ and to each other?

  b. In general, how is the graph of $k(x) = \log_a x$ where $a > 1$ related to the graph of $f(x) = \log x$? Explain your reasoning.

In Unit 1, "Rates of Change," you probably discovered that the derivative of an exponential function is an exponential function. It should seem natural to inquire about the nature of the derivative of a logarithmic function. Recall that the derivative of a function can be approximated by the difference quotient:

$$f'(x) \approx \frac{f(x + h) - f(x)}{h}$$

3. Now look more closely at the rate of change of $f(x) = \log x$.

  a. Write the difference quotient for $f(x) = \log x$ using $h = 0.0001$. Store it in Y_1 on your calculator.

  b. Approximate $f'(x)$ at $x = 1, 2, 3, 4,$ and 5. Then, use the **Table** function on your calculator to make a table of values of the derivative for these values and for additional integral and nonintegral values of x.

  c. Add a second column to your table by setting $Y_2 = XY_1$. What do you notice about the values of Y_2? Conjecture a rule for calculating $f'(x)$ for all $x > 0$.

4. The numerical pattern you observed in Activity 3 suggested a possible rule for the derivative of $f(x) = \log x$. Now consider the difference quotient for the log function more generally.

  a. Use your knowledge of logarithms and algebra to provide reasons for each step in the reasoning process below.

$$f'(x) \approx \frac{\log(x + h) - \log(x)}{h}$$
$$= \frac{\log\left(\frac{x + h}{x}\right)}{h}$$
$$= \frac{1}{h} \cdot \log\left(1 + \frac{h}{x}\right)$$
$$= \frac{1}{x} \cdot \frac{x}{h} \cdot \log\left(1 + \frac{h}{x}\right)$$
$$= \frac{1}{x} \cdot \log\left(1 + \frac{h}{x}\right)^{\frac{x}{h}}$$

  b. In Activity 3 Part c, you observed that $f'(x) \approx \frac{0.43429}{x}$. What does this allow you to conclude about $\log\left(1 + \frac{h}{x}\right)^{\frac{x}{h}}$?

  c. Find an approximate value for a number whose logarithm is 0.43429.

Like π, the value you obtained in Part c of Activity 4 is a special irrational number. It is called "e," and is approximated by 2.71828182846. The number e arises naturally when computing instantaneous rates of change of logarithmic functions.

5. The number e also arises when interest is compounded *continuously.* Recall that the balance in a bank account with an initial deposit of $1,000 at 3% annual interest will be $1,000(1.03) = $1,030 after one year. If the bank applied $\frac{1}{12}$ of the interest rate to your balance each month and compounded monthly, the value of the account after one year would be

$$1,000 \cdot \left(1 + \frac{0.03}{12}\right)^{12} \approx 1,000 \cdot (1.030416)$$

or $1,030.42. Here 3% is the *nominal* annual interest rate, whereas the *effective* annual rate when compounded monthly is 3.042%.

a. Investigate the effect of compounding a 3% nominal annual rate more frequently. What are the effective annual rates if interest is compounded daily? Hourly? Every minute?

b. If the 3% nominal rate were compounded continuously, the effective rate would be approximately 3.0454534%. Your balance at the end of t years would be $1,000 \cdot (1.030454534)^t$. Express 1.030454534 as a power with the nominal rate of 0.03 as exponent. What is the base? Write an equation that gives the value of your account in t years and in which the nominal rate appears.

c. It can be shown that, in general, a nominal interest rate of r will give an effective interest rate of $e^r - 1$. Suppose a 6% annual rate is applied to $1,000 continuously. Find the effective annual rate.

d. Suppose an effective annual rate is 23.3678%. Find the nominal annual rate that is being compounded continuously.

Since any positive number other than 1 can be the base of a logarithm, so too can e.

$$y = \log_e x \text{ if and only if } e^y = x.$$

The base e logarithm is called the **natural logarithm** and denoted "ln." (ln x represents $\log_e x$.)

6. Because natural logs are widely used in mathematics and its applications, the ln function, like the log function, is a built-in function on most calculators. Use your calculator to help you investigate and describe characteristics of the natural log function.

a. On the same axes, graph $y = \ln x$ and $y = e^x$. How are these graphs related? How are the functions related?

b. What are the domain and range of $y = \ln x$?

c. Describe the pattern of change shown in the graph of $y = \ln x$.

d. Where does the graph of $y = \ln x$ cross the x-axis? Express this fact symbolically. Explain why the graph does not cross the y-axis.

e. For what values of x is $\ln x$ positive? Negative?

f. Find $\ln 0.5$ and $\ln 20$.

7. How are the graphs of $f(x) = \ln x$ and $g(x) = \log x$ related? Use the change of base formula to express the relationship symbolically.

8. Now investigate the derivative of $f(x) = \ln x$.

a. Make a table similar to the one in Activity 3. What do you notice about $xf'(x)$?

b. Write a rule for $f'(x)$ for all $x > 0$. Sketch the graph of $y = f'(x)$.

9. In Unit 3, you proved properties for common logarithms that were similar to properties for exponents. Restate those properties in terms of natural logarithms. Explain why these properties hold for logarithms with base e.

10. Use natural logarithms to solve each equation for x.

a. $5.2^x = 28$

b. $a^{2x} = 31, a > 0$

c. $a^x = b, a > 0, b > 0$

d. $7^{x-3} = 41$

e. $2.5(1.73)^{3x-2} = 5.82$

Checkpoint

In this investigation, you considered logarithms with bases other than 10.

a Compare natural logarithms and common logarithms.

b Describe how either natural or common logarithms can be used to compute values of logarithms with other bases.

c Explain why properties of common and natural logarithms also hold for logarithms with base b, $b > 0$, $b \neq 1$.

d Compare the instantaneous rates of change at $x = a$ for $f(x) = \log x$ and $g(x) = \ln x$.

Be prepared to share your ideas with the entire class.

Think about the similarities between common logarithms and natural logarithms as you complete these tasks.

a. Evaluate $\log_{4.6} 21$ using common logarithms and using natural logarithms. Compare the results.

b. Find the instantaneous rate of change of $f(x) = \log x$ at $x = 3.5$. Find the instantaneous rate of change of $g(x) = \ln x$ at $x = 3.5$.

c. Without computing, make a conjecture as to the relation between $\ln 10$ and $\log e$. Find $\ln 10$ and $\log e$. Was your conjecture correct?

INVESTIGATION 4 Solving Logarithmic Equations

In Unit 3, "Logarithmic Functions and Data Models," you learned how to use data transformations to check if a set of data points is modeled reasonably well by an exponential function or by a power function. For example, the context or pattern of data in a scatterplot may suggest that the data *might* follow an exponential model (or a power model). If the pattern in the (x, y) data is *exponential*, then the plot of $(x, log\ y)$ will be approximately linear. If the pattern in the (x, y) data is modeled well by a *power function*, then the plot of $(log\ x, log\ y)$ will be approximately linear.

1. You can use properties of logarithms and symbolic reasoning to explain why these data transformations permit you to draw valid conclusions about the underlying models.

   a. Show that $y = ab^x$ is equivalent to $\log y = \log a + x \log b$ for $a > 0$, $b > 0$.
      - Explain why the logarithmic equation is linear in form.
      - What is the slope of the line? The vertical intercept?
      - Describe the coordinates of the points on the graph of the logarithmic equation.

   b. Show that $y = ax^b$ is equivalent to $\log y = \log a + b \log x$ for $a > 0$, $x > 0$.
      - Explain why the logarithmic equation is linear in form.
      - What is the slope of the line? The vertical intercept?
      - Describe the coordinates of the points on the graph of the logarithmic equation.

2. Look back at the data (page 436) on the average distance from the Sun to a planet and the number of Earth days needed for that planet to orbit the Sun. The linear regression equation of the transformed data is

$$\log t = 1.50 \log d - 0.39.$$

   a. Explain why this equation is linear in $\log t$ and $\log d$.

b. What does the fact that the log-log transformation linearizes the original data tell you?

c. When Janelle and Dennis were asked to re-express $\log t = 1.50 \log d - 0.39$ with t as a function of d, they reasoned as shown in the following set of equations. Justify each step in their derivation of the modeling equation, or identify where their reasoning is flawed.

$$\log t = 1.50 \log d - 0.39$$
$$0.39 = 1.50 \log d - \log t$$
$$0.39 = \log d^{1.50} - \log t$$
$$0.39 = \log \left(\frac{d^{1.50}}{t} \right)$$
$$10^{0.39} = \frac{d^{1.50}}{t}$$
$$t = \frac{d^{1.50}}{10^{0.39}} = 10^{-0.39} \cdot d^{1.50} \text{ or } t = 0.4074 \cdot d^{1.50}$$

d. Is $t = 0.4074 \cdot d^{1.50}$ a power function? Why or why not?

3. If you transform the (*average distance, orbital time*) data using the natural logarithm, you will get the linear regression equation:

$$\ln t = 1.50 \ln d - 0.8954$$

a. Why is it reasonable that this equation will be different from the linear regression equation obtained when the data were transformed using the common logarithm?

b. Transform the above equation into an equation expressing t as a function of d.

c. Notice that you get the same model when the data are transformed by common or natural logarithms. Explain why this is reasonable.

4. Solve each equation for y.

a. $\log y = a \log x + b_1$

b. $\ln y = a \ln x + b_2$

5. In 1905, when the study of radioactivity was just beginning, Meyer and von-Schwiedler reported data giving the time in days and the relative radioactivity for a radioactive substance. (See Unit 3, pages 190–191). If you use the natural logarithm to transform the radioactivity data, a plot of (*time, ln relative activity*) data will be linear. The linear regression equation is $\ln y = -0.0752x + 3.3$.

a. Re-express $\ln y = -0.0752x + 3.3$ in a form with no logarithms.

b. When common logarithms are used, the regression equation is $\log y = -0.03264x + 1.433$. Transform this equation into a form with no logarithms.

c. Compare your equations in Parts a and b.

d. Write a general rule expressing y as a function of x when the linear regression equation of transformed data is $\log y = ax + b$. When it is $\ln y = cx + d$.

In the previous activities, you may have observed that equations involving logarithms are not as easy to interpret as are equivalent forms involving no logarithms. Transforming equations involving logarithms to equivalent equations without logarithms is a general strategy to determine the nature of an underlying relationship, or, in the case of a single variable, to solve an equation.

6. Use the properties of logarithms and symbolic reasoning to express y as a function of x.

   a. $\log y = \log (x - 1) + 1$

   b. $\log y = -\log (x + 1) + 1$

   c. $\log y = 1 - 2 \log (x + 3)$

   d. $\ln y = 2 - 2 \ln (x - 1)$

   e. $\ln y = k \ln (3x - 1) + 1$

   f. $\log y = 3x + 1$

   g. $\log y = 2x - 1$

   h. $\ln y = 0.5x + 1$

7. Use symbolic reasoning to solve each equation for x. Check your solutions by substitution in the original equation. If no solution exists, explain why.

   a. $\log (x + 1) - \log (x - 1) = 2$

   b. $\ln (x + 1) + \ln (x - 1) = 0$

   c. $\ln (3x - 1) - \ln (x + 1) = 0$

   d. $\log (2x + 1) - 2 \log (x - 1) = 1$

Checkpoint

Linearizing data to reveal underlying mathematical relationships requires an understanding of statistical methods and basic function models, as well as skill in symbolic reasoning.

a The log and the log-log transformations of (x, y) data often lead to linear patterns in $(x, log\ y)$, $(x, ln\ y)$, $(log\ x, log\ y)$, or $(ln\ x, ln\ y)$. If each linearized log model is re-expressed with y as a function of x, what kind of function do you get?

b Given a functional relationship expressed in logarithms, how would you transform it into a "$y = ...$" relationship or solve it for a variable?

Be prepared to explain your ideas and reasoning to the class.

On Your Own

Each equation below is the regression equation for a different set of transformed data. In each case, write an equation expressing y as a function of x.

a. $\ln y = 0.258x - 0.713$

b. $\log y = 1.73 \log x + 0.159$

MORE

Modeling

1. The velocity at which a diver enters the water is related to the height of the dive. The following table gives the entry velocities for five heights.

Height (in feet)	10	15	25	50	100
Entry Velocity (in mph)	17.2	21.2	27.4	38.6	54.5

Royal Gorge

  a. Based on the context, would you expect the data to be linear? Explain. What function model do you think might fit these data well?

  b. Does the log transformation (*height, ln velocity*) straighten the data?

  c. Does the log-log transformation (*ln height, ln velocity*) straighten the data?

  d. Write a linear regression equation using the transformed data that are most nearly linearly related.

  e. Write an equation expressing velocity as a function of the height.

  f. Suppose the entry velocity of a cliff diver was recorded at 30 mph. How high was the cliff?

  g. How is the entry velocity affected by doubling the height of the dive?

  h. The highest suspension bridge in the world is the bridge over the Royal Gorge of the Arkansas River in Colorado. It is 1,053 feet above the water. What is the entry velocity of an object dropped off this bridge?

2. Suppose that $f(t) = 100\left(16^{\frac{t}{4}}\right)$ models the growth of a colony of bacteria in a laboratory study.

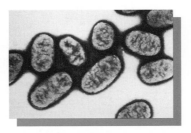

  a. Find the initial count of the bacteria in the colony.

  b. If *t* is measured in weeks, find the growth factor.

  c. What is the doubling time for this bacteria?

  d. Write an equivalent function rule for $f(t)$ with base 2.

  e. Write an equivalent function rule for $f(t)$ with base 10.

  f. Express $f(t)$ as an exponential function with base *e*.

3. Suppose that the decay of a radioactive substance is modeled by the function $h(x) = 100(0.125^{\frac{x}{3}})$, where x is measured in years and $h(x)$ is measured in grams.

   a. How much of the substance is there initially?

   b. What is the decay factor?

   c. Express $h(x)$ as an exponential function with exponent x.

   d. Express $h(x)$ as an exponential function with base e.

   e. Express $h(x)$ as an exponential function with base 10.

   f. Are $25(2^{2-x})$ and $100(0.125^{\frac{x}{3}})$ equivalent expressions? Justify your answer.

4. The following table gives the population per square mile of land in the United States for 10-year periods from 1870 through 2000.

Year	1870	1880	1890	1900	1910	1920	1930	1940	1950	1960	1970	1980	1990	2000
People Per Square Mile	10.9	14.2	17.8	21.5	26.0	29.9	34.7	37.2	42.6	50.6	57.4	64.0	70.3	79.6

Source: *The World Almanac and Book of Facts*, 2000, p. 386; www.factfinder.census.gov/servlet/BasicFactsServlet

   a. Use logarithmic transformations to find a model for the population density as a function of year.

   b. Use your model and symbolic reasoning to predict the year in which the number of people per square mile first exceeded 60.

   c. What does your model predict for the year 2010?

5. Suppose you wish to invest $1,000 in a savings account and that your two best options are a 5% annual nominal rate that is compounded quarterly or a 4.7% annual nominal rate that is compounded continuously. Which is the better investment? Explain.

Organizing

1. Use symbolic reasoning to solve each exponential equation.

   a. $8^x = 32$

   b. $2^{-x} = 16$

   c. $3^x = 17$

   d. $5(2^x) = 12$

   e. $6^{0.3x} = 17$

   f. $8^{\frac{x}{4}} = 32$

   g. $4^{x-2} = 32$

   h. $2.7(4.1^{3x-2}) = 8.6$

   i. $e^x = 3x + 5$

2. Solve each equation for x. If no solution exists, explain why.

   a. $\log (x + 1) - \log (x - 3) = 3$

   b. $\log (2x - 1) - \log x = 1$

   c. $[\ln (x - 1)]^2 = 2$

   d. $\ln (2x - 2) - \ln (x - 1) = \ln x$

3. Suppose that the linear equation $\ln y = rx + s$ is equivalent to the exponential equation $y = a(b^x)$. Write expressions for r and s in terms of a and b.

4. Consider the exponential function $f(x) = a(b^x)$, $a > 0$, $a \neq 1$, $b > 0$, $b \neq 1$.

   a. Express $f(x)$ as an exponential function with base 10.

   b. Express $f(x)$ as an exponential function with base e.

   c. Express $f(x)$ as an exponential function with base a.

5. Consider the graphs of $y = b^x$ for $b = 10$, $b = e$, and $b = 2$, where $x > 0$.

   a. Do any of these graphs intersect the line $y = x$? If not, which graph comes closest to intersecting the line?

   b. Use graphical or numerical reasoning to approximate the largest value of b less than 2 for which the graphs of $y = b^x$ and $y = x$ intersect. Estimate the coordinates of the point of intersection.

   c. The x-coordinate of the intersection point of the graphs in Part b may look familiar. It is approximately e. Use this observation to find the exact largest value of b for which the graphs intersect.

Reflecting

1. In what sense does a common logarithm "undo" the effects of the exponential expression 10^x?

2. How is solving a linear equation in x such as $ax + b = c$ similar to, and different from, solving a linear equation in $\log x$ such as $a \log x + b = c$?

3. Common logarithms use 10 as a base and natural logarithms use e as a base. How can you convert the common logarithm of a number into the natural logarithm and vice versa?

4. Think about the relationship between exponents and logarithms.

   a. What is the value of $e^{\ln x}$? Explain your reasoning and specify the values of x for which the relationship holds.

   b. What is the value of $\ln e^x$? Explain your reasoning and specify the values of x for which the relationship holds.

5. Standard Federal Bank currently offers a 36-month certificate of deposit that pays interest at the nominal rate of 6% per year. Study the table below showing the effective yield for various compounding periods.

Compounding Frequency	Effective Yield
annual	1.06
monthly	1.0616778
daily	1.0618313
hourly	1.0618363
⋮	⋮

As the compounding periods grow shorter, the effective yield appears to converge to a particular number. What is that number?

Extending

1. In Course 3, Unit 7, "Discrete Models of Change," you investigated *finite* series. Although both π and e are irrational numbers, each can be defined as an *infinite* series of rational numbers:

$$\pi = 4\sum_{n=0}^{\infty} \frac{(-1)^n}{2n+1} \qquad e = \sum_{n=0}^{\infty} \frac{1}{n!}$$

   a. Write the first five terms in each of these series.

   b. How many terms of the series would be needed to write the values of π and e correct to two decimal places? Which series requires fewer terms to get an accurate approximation?

   c. How many terms of the series representation of e are needed to produce the decimal approximation of e equal to that displayed by your calculator?

2. In Investigation 3, the number e arose in the context of considering a savings account in which interest was compounded continuously. The constant e can also be found by examining the derivative of exponential functions. Recall that if $f(x) = b^x$, then the derivative of $f(x)$ is approximated by the difference quotient $\frac{f(x+h) - f(x)}{h}$ or $\frac{b^{x+h} - b^x}{h}$ for small values of h.

   a. Show that $f'(x)$ can be approximated by $k \cdot b^x$. Write an expression for k.

   b. Compute values of k when h is small, for example, 0.000001, and b takes on integral values 1, 2, 3, … .

   c. Plot the points (b, k). To which family of functions does the graph of the (b, k) points seem to belong?

   d. Find the value b for which $k = 1$. Does this number look familiar?

   e. If $g(x) = e^x$, write an expression for $g'(x)$.

   f. What is the instantaneous rate of change for $f(x) = b^x$, $b > 0$?

3. The equation for the value of an account with initial deposit A compounded continuously for t years with a nominal continuous annual rate of k is $V = Ae^{kt}$.

   a. If $k = 5.5\%$ and $A = \$100$, find V when $t = 5$ and when $t = 20$.

   b. Find the effective annual rate of return.

   c. How many years will it be before the original investment doubles?

   d. Rewrite $V = 100e^{0.055t}$ in the form $V = AB^t$.

   e. If $V = 100e^{0.08t}$, then $V = 100(1.08329^t)$. What does 0.08329 represent?

4. As a rule of thumb, an investment earning an annual interest rate of r percent will double in approximately $\frac{72}{r}$ years. This is called the *Rule of 72*.

   a. How well does the Rule of 72 work for rates of 1, 2, 6, 8, and 12 percent using annual compounding?

   b. Using the formula for continuous compounding with rate k, $V = Ae^{kt}$, find an expression for the doubling time for V.

   c. Use your result in Part b to compute the doubling time for $k = 1, 2, 6, 8,$ and 12 percent.

   d. Where did the "72" come from in the "Rule of 72"?

5. In Extending Task 2, you sought the value of b for which $\frac{b^h - 1}{h} \approx 1$ when h was small.

   a. Solve $\frac{b^h - 1}{h} = 1$ for b and then estimate the value of b.

   b. Estimate the value of b by substituting $h = \frac{1}{n}$ in the equation $b = (1 + h)^{\frac{1}{h}}$.

6. Solve each equation for x. Identify the conditions that must be placed on a, b, and c so that a solution exists.

   a. $a \cdot b^x = c$

   b. $a \cdot b^{kx} = c$

   c. $a \cdot b^{x + r} = c$

   d. $a \cdot b^{kx + r} = c$

1. How much will a jacket that normally sells for $180, not including tax, cost during a 30 percent off sale in a state whose sales tax is 5 percent?

    (a) $117.00   (b) $132.30   (c) $142.50   (d) $119.70   (e) $155.00

2. Simplify: $\left(\dfrac{(x+2)^2}{x^2-4}\right)\left(\dfrac{x-2}{x+2}\right)$

    (a) $x+2$ (b) $\dfrac{x+2}{x-2}$ (c) $\dfrac{1}{x+2}$ (d) 1 (e) $\dfrac{x-2}{x+2}$

3. The formula $F = \dfrac{9}{5}C + 32$ expresses degrees Fahrenheit (F) in terms of degrees Celsius (C). What is C when $F = 38$?

    (a) $\dfrac{342}{5}$ (b) $-\dfrac{98}{9}$ (c) $\dfrac{10}{3}$ (d) $\dfrac{502}{5}$ (e) $\dfrac{350}{9}$

4. If $f(x) = x^2 + x - 3$, then $f(-2a) =$

    (a) $4a^2 + 2a - 3$ (b) $a^2 + 2a - 3$ (c) $-4a^2 - 2a - 3$

    (d) $4a^2 + 2a + 3$ (e) $4a^2 - 2a - 3$

5. Solve for x: $x(x - 4) = 5$

    (a) $x = 5$ or $x = -1$ (b) $x = 5$ or $x = 9$ (c) $x = 5$ or $x = 1$

    (d) $x = 0$ or $x = 4$ (e) $x = 3$ or $x = -3$

6. Which best represents the graph of the solution set of $x^2 + 2x < 0$?

7. The graph of the function $y = f(x)$ is shown. Which of the following is a possible algebraic model representing the function? The scale is 1 on both axes.

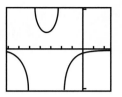

(a) $f(x) = \dfrac{1}{(x+5)(x+2)}$

(b) $f(x) = \dfrac{1}{(x-5)(x-2)}$

(c) $f(x) = \dfrac{-1}{(x+5)(x+2)}$

(d) $f(x) = \dfrac{-1}{(x-5)(x-2)}$

(e) $f(x) = \dfrac{-1}{(x-5)(x+2)}$

8. The lengths of the sides of a rectangle are x and y. If the length of each side is doubled, the new area is

(a) $8xy$ (b) $2xy$ (c) $2x + 2y$ (d) $4x + 4y$ (e) $4xy$

9. Which of the following is equivalent to $2 \log a + \log b - \log c$?

(a) $\log (a^2 + b - c)$ (b) $\log (2a + b - c)$ (c) $2 \log \left(\dfrac{ab}{c}\right)$

(d) $\log \left(\dfrac{a^2 b}{c}\right)$ (e) $\log a^2 bc$

10. This is the graph of which of the following equations?

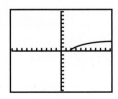

(a) $y^2 = x - 2$ (b) $y = \sqrt{x+2}$ (c) $y^2 = x + 2$

(d) $y = \sqrt{x} + 2$ (e) $y = \sqrt{x-2}$

Lesson 2

Reasoning with Trigonometric Functions

In Unit 2, "Modeling Motion," you learned that projectile motion such as the path of a baseball or golf ball is the sum of a horizontal and a vertical component. When the initial velocity is V and the angle of elevation is θ, the initial magnitudes of the components are $V \cos \theta$ and $V \sin \theta$.

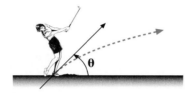
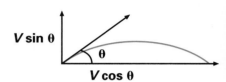

Ignoring air resistance, the height (in feet) of the projectile after t seconds is given by $h(t) = (V \sin \theta) \cdot t + (16t \cdot t \sin 270°)$ or $h(t) = (V \sin \theta) \cdot t - 16t^2$. While in the air, the horizontal distance (in feet) the projectile travels in t seconds is given by $d(t) = (V \cos \theta) \cdot t$.

Athletes are often interested in the value of $d(t)$ when $h(t) = 0$. That is, they are interested in how far the baseball or golf ball will travel before it returns to the ground. This distance (in feet), called the *range*, is given by

$$R(\theta) = \frac{V^2 \cos \theta \sin \theta}{16}$$

where V is in feet per second.

Think About This Situation

Consider a golfer who consistently hits the ball so that it leaves the tee with an initial velocity of about 110 ft/sec.

a Explain why $h(t) = 0$ when $t = 0$ and when $t = \frac{V \sin \theta}{16}$.

b What is the range of the golf ball when hit at an angle of 30° to the ground?

c To get a longer hit than that in Part b, should the golfer increase or decrease the angle of the hit? Why?

d What angle of elevation do you think would produce the longest hit? Explain your reasoning.

e Describe the graph of $R(\theta)$ for a fixed initial velocity of 110 ft/sec.

f Compare the graphs of $R(\theta)$ and $f(\theta) = \sin \theta$.

The range function $R(\theta)$ involves the product of two trigonometric functions. This product can be expressed in an equivalent form as a single trigonometric function with which you are familiar. In this lesson, you will develop key trigonometric *identities* which are useful in making such transformations.

INVESTIGATION 1 Extending the Family of Trigonometric Functions

In *Contemporary Mathematics in Context*, Course 2, the trigonometric functions, sine, cosine, and tangent, were defined as ratios of the lengths of sides of right triangles. Later the definitions for sine and cosine were extended to permit angles greater than 90° and angles less than 0°. These ideas and the definition of the tangent function are formalized below.

Consider any point $A(x, y)$, other than the origin, in a coordinate plane. Let θ be the measure of the angle determined by rotating a position vector of length r from a position along the positive x-axis to its position $\overrightarrow{OA}$. If the rotation is counter-clockwise, θ is positive; if clockwise, θ is negative. The trigonometric functions are functions of θ and are defined in terms of x, y, and r where $r = \sqrt{x^2 + y^2}$, the length of $\overrightarrow{OA}$, as follows:

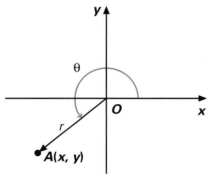

$$\sin \theta = \tfrac{y}{r} \qquad \cos \theta = \tfrac{x}{r} \qquad \tan \theta = \tfrac{y}{x}, x \neq 0$$

For convenience, θ will be used to denote both an angle and its measure. It will be clear from the context which meaning is intended.

1. For this activity, assume that θ is an angle in **standard position** in a coordinate plane with its vertex at the origin and its *initial side* along the positive x-axis, and that $A(x, y)$ is a point on the *terminal side* of the angle.

   a. Sketch θ and find $\sin \theta$, $\cos \theta$, and $\tan \theta$ for $A(x, y)$ when the coordinates of point A are as indicated.

   i. $(3, -2)$ **ii.** $(-4, -3)$ **iii.** $(2, 5)$ **iv.** $(-3, 4)$

   b. Examine your results in Part a. Whether the value of a trigonometric function is positive or negative depends on the quadrant in which the terminal side of θ lies. For each quadrant in which the terminal side of an angle may lie, determine

   ■ the corresponding range of angle measures for θ; and

   ■ whether $\sin \theta$, $\cos \theta$, and $\tan \theta$ are positive or negative.

   Summarize your findings in a chart similar to the one on the next page.

Quadrant II	Quadrant I
$90° < \theta <$ ___ $°$	$0° < \theta < 90°$
___ $< \theta <$ ___ radians	$0 < \theta < \frac{\pi}{2}$ radians
$\sin \theta > 0$	$\sin \theta > 0$
$\cos \theta$ ___ 0	$\cos \theta > 0$
$\tan \theta$ ___ 0	$\tan \theta > 0$
Quadrant III	Quadrant IV
___ $° < \theta <$ ___ $°$	___ $° < \theta < 360°$
___ $< \theta <$ ___ radians	___ $< \theta < 2\pi$ radians
$\sin \theta$ ___ 0	$\sin \theta$ ___ 0
$\cos \theta$ ___ 0	$\cos \theta$ ___ 0
$\tan \theta$ ___ 0	$\tan \theta$ ___ 0

c. The chart in Part b does not consider the values of $\sin \theta$, $\cos \theta$, and $\tan \theta$ when the terminal side of θ is on the *x*- or *y*-axis.

- For what values of θ is its terminal side on the *x*-axis? What can you say about values of $\sin \theta$, $\cos \theta$, and $\tan \theta$ in these cases?

- For what values of θ is its terminal side on the *y*-axis? What can you say about values of $\sin \theta$, $\cos \theta$, and $\tan \theta$ in these cases?

2. Now suppose $A(x, y)$ is a point, other than the origin, on the terminal side of an angle θ in standard position, and $\sqrt{x^2 + y^2} = r$.

a. Find the coordinates of points A', A'', and A''' where:

- Point A' is the image of point A reflected across the *x*-axis.

- Point A'' is the image of point A reflected across the *y*-axis.

- Point A''' is the image of point A rotated $180°$ about the origin.

Why are points A', A'', and A''' on the circle with center O and radius r?

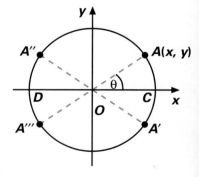

b. Assuming each angle is in standard position and $m\angle AOC = \theta$, express $m\angle A'OC$, $m\angle A''OC$, and $m\angle A'''OC$ in terms of θ.

c. Investigate how knowing the value of one of the trigonometric functions of an angle θ permits you to find values of the other two trigonometric functions. Assume the terminal side of θ is in the first quadrant.

- If $\sin \theta = \frac{3}{5}$, what other angle (in a different quadrant) has the same sine? Find the cosine and tangent for both angles.

- If $\cos \theta = \frac{4}{5}$, what other angle has the same cosine? Find the sine and tangent for both angles.

- If $\tan \theta = \frac{3}{4}$, what other angle has the same tangent? Find the sine and cosine for both angles.

d. Use reasoning similar to that in Part c to complete the following:

- If $\sin \theta = -\frac{3}{5}$ and θ is in quadrant III, what other angle has the same sine? Find the cosine and tangent for both angles.

- If $\cos \theta = -\frac{4}{5}$ and θ is in quadrant II, what other angle has the same cosine? Find the sine and tangent for both angles.

- If $\tan \theta = -\frac{3}{4}$ and θ is in quadrant II, what other angle has the same tangent? Find the sine and cosine for both angles.

For any given position vector with length r, the terminal point of the vector is always on a circle of radius r. Thus, the coordinates of the terminal point are functions of the cosine and sine of the direction angle of the vector: $x = r \cos \theta$ and $y = r \sin \theta$.

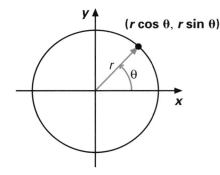 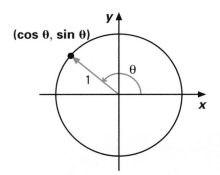

In the case of a **unit circle**, a circle with radius 1 as above on the right, the coordinates are simply $x = \cos \theta$ and $y = \sin \theta$, and $\tan \theta = \frac{y}{x} = \frac{\sin \theta}{\cos \theta}$, $\cos \theta \neq 0$.

3. The relationship between points on a unit circle and the values of the cosine and sine functions can be used to help sketch graphs of the trigonometric functions.

a. Sketch graphs of $y = \cos \theta$ and $y = \sin \theta$ for $-2\pi \leq \theta \leq 2\pi$ by imagining the terminal point of a position vector rotating about a unit circle. Check your graphs against calculator- or computer-produced graphs and resolve any differences.

b. Sketch a graph of $y = \tan \theta$ for $-2\pi \leq \theta \leq 2\pi$ by reasoning with the relationship $\tan \theta = \frac{\sin \theta}{\cos \theta}$ and your graphs of the sine and cosine functions. Compare your sketch with a calculator- or computer-produced graph and resolve any differences.

c. How is the information you gathered in Activity 1 Parts b and c related to the graphs?

d. For each of the three trigonometric functions:

  ■ state the domain and range;

  ■ state the period.

4. You can use the graphs you produced in Activity 3 to aid in reasoning about the solutions of trigonometric equations.

   a. How many values of θ satisfy the equation $\cos \theta = \frac{3}{5}$, where $0 \leq \theta < 2\pi$ radians? Explain your reasoning.

   b. Use the inverse cosine (**cos**⁻¹) function on your graphing calculator to find a value of θ so that $\cos \theta = \frac{3}{5}$. Find any other values of θ between 0 and 2π satisfying this equation.

   c. Solve each equation below for all values of θ, $0 \leq \theta < 2\pi$.

   i. $\sin \theta = \frac{1}{2}$ **ii.** $\cos \theta = -0.7$
   iii. $\tan \theta = 0.5\sqrt{2}$ **iv.** $\sin \theta = 2$

   d. Explain why, if a trigonometric equation involving $\sin \theta$, $\cos \theta$, or $\tan \theta$ has a solution, then there is usually a second solution on the interval $0 \leq \theta < 2\pi$.

   e. What types of trigonometric equations might have more than two solutions on the interval $0 \leq \theta < 2\pi$? Fewer than two solutions?

The reciprocals of the three trigonometric functions, $\sin \theta$, $\cos \theta$, and $\tan \theta$, give a new set of trigonometric functions called the **reciprocal functions**.

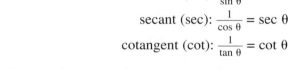

$$\text{cosecant (csc): } \frac{1}{\sin \theta} = \csc \theta$$

$$\text{secant (sec): } \frac{1}{\cos \theta} = \sec \theta$$

$$\text{cotangent (cot): } \frac{1}{\tan \theta} = \cot \theta$$

5. Explore the graph of $y = \sec \theta$ for $-2\pi \leq \theta \leq 2\pi$.

   a. Make a sketch of the graph of $y = \cos \theta$ and of its reciprocal function $y = \sec \theta$ on the same set of axes. Identify common points.

   b. Identify the values of θ, $-2\pi \leq \theta \leq 2\pi$, for which $y = \sec \theta$ is not defined. How is this revealed in the definition of the function? How is it seen in the graph of the function?

   c. Describe the range of $y = \sec \theta$.

   d. Is the function $y = \sec \theta$ periodic? If so, give the period. Why does this make sense in terms of the definition of the function?

   e. Explain how you can quickly sketch the graph of $y = \sec \theta$ by remembering the shape of, and key points on, the graph of $y = \cos \theta$.

6. Repeat Activity 5 for the function $y = \sin \theta$ and its reciprocal function $y = \csc \theta$.

7. Repeat Activity 5 for the function $y = \tan \theta$ and its reciprocal function $y = \cot \theta$.

8. Investigate how reasoning with the reciprocal trigonometric functions is similar to reasoning with the trigonometric functions themselves.

   a. Sketch a right triangle with sec $\theta = 2$. Evaluate the remaining five trigonometric functions of θ.

   b. Sketch two position vectors such that csc $\theta = 2$. Find the remaining trigonometric function values for the direction angle of each vector.

   c. Sketch two position vectors such that cot $\theta = -1$. Find the remaining trigonometric function values for the direction angle of each vector.

   d. Find all solutions to each equation below on the interval $0 \leq \theta < 2\pi$. Compare the procedures you used with those used by others.

   i. csc $\theta = 4$ **ii.** sec $\theta = -2$ **iii.** cot $\theta = \sqrt{3}$

Checkpoint

Suppose $A(x, y)$ is a point on the terminal side of an angle θ in standard position and r is the distance from point A to the origin.

ⓐ Write the definitions of the six trigonometric functions in terms of x, y, and r.

ⓑ For what values of x and y are each of the six trigonometric functions not defined? What values of θ, $-2\pi \leq \theta \leq 2\pi$, are associated with those values of x and y? What values of θ, $-360° \leq \theta \leq 360°$ are associated with those values of x and y?

ⓒ Explain how sketches of the graphs of $y = \sec \theta$, $y = \csc \theta$, and $y = \cot \theta$ can be obtained from examination of the graphs of $y = \cos \theta$, $y = \sin \theta$, and $y = \tan \theta$, respectively.

Be prepared to compare your definitions of the trigonometric functions and their domains and your graphing methods to those of other groups.

▶On Your Own

Use symbolic reasoning to complete the following tasks.

a. Suppose $\sin \theta = -\frac{5}{13}$ and $0 \leq \theta < 2\pi$. How many solutions are there to this equation? Find all the solutions.

b. Suppose $\sin \theta = \frac{12}{13}$ and $\cos \theta = -\frac{5}{13}$. How many values of θ between 0 and 2π satisfy both equations? Find all such θ.

c. Using the definitions of the trigonometric functions recorded at the Checkpoint:

   ▪ Prove that $\tan \theta = \frac{\sin \theta}{\cos \theta}$, $\cos \theta \neq 0$.

   ▪ Prove that $\cot \theta = \frac{\cos \theta}{\sin \theta}$, $\sin \theta \neq 0$.

INVESTIGATION 2 Proving Trigonometric Identities

One important aspect of trigonometry is the use of many formulas expressing interrelationships among the trigonometric functions. The quotient formulas $\tan \theta = \frac{\sin \theta}{\cos \theta}$ and $\cot \theta = \frac{\cos \theta}{\sin \theta}$ are examples of *trigonometric identities*. An **identity** is a statement that is true for all replacements of a variable for which the statement is defined. You have seen other identities in your previous work in algebra. For example, $a \cdot \frac{1}{a} = 1$ is an identity for all nonzero real numbers and $\log (ab) = \log a + \log b$ is an identity for all positive real numbers. Examples of identities involving all real numbers are $a(b + c) = ab + ac$ and $(2x + 1)^2 = 4x^2 + 4x + 1$.

Identities are useful because they permit you to rewrite expressions so that they are more recognizable and often easier to interpret or work with in solving equations. Consider again the rule given at the beginning of this lesson (page 458) for the number of feet that a projectile will travel.

$$R(\theta) = \frac{V^2 \cos \theta \sin \theta}{16}$$

For a fixed value of V, R is a function of θ, the angle of the projectile.

The fact that $R(\theta)$ involves the product "$\cos \theta \sin \theta$" makes it difficult to visualize the shape of its graph or where it may have a maximum or minimum. In Modeling Task 1 (page 472), you will establish the identity $2 \cos \theta \sin \theta = \sin 2\theta$, or, equivalently, $\cos \theta \sin \theta = \frac{\sin 2\theta}{2}$. Using this identity you can rewrite $R(\theta)$ as $\frac{V^2 \sin 2\theta}{32}$, which is an easier form to interpret.

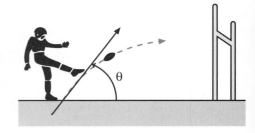

There are many additional useful identities involving trigonometric functions. You will examine several of the basic identities in this investigation, and you will learn how to prove that a statement is an identity.

1. A key identity you encountered in your previous study is the statement $(\sin \theta)^2 + (\cos \theta)^2 = 1$. For simplicity, $(\sin \theta)^2$ is often written $\sin^2 \theta$.

   a. Describe the graph of $f(\theta) = \sin^2 \theta + \cos^2 \theta$.

   b. $\sin^2 \theta + \cos^2 \theta = 1$ is called a *Pythagorean identity*. Prove this identity using one of the diagrams on page 461. Explain why the name of this identity is appropriate.

   c. Is $\sec^2 \theta - \tan^2 \theta = 1$ an identity?

   - If so, use the definitions of secant and tangent to prove it. If not, provide a counterexample.

   - Prove it beginning with the identity $\sin^2 \theta + \cos^2 \theta = 1$.

d. Is $\tan^2 \theta + \sec^2 \theta = 1$ an identity? If so, prove it. If not, explain how you know.

e. Is $\csc^2 \theta - \cot^2 \theta = 1$ an identity? If so, prove it. If not, explain how you know.

You now have at your disposal eight fundamental relationships that you can use to rewrite trigonometric expressions in equivalent forms.

Fundamental Relationships

Reciprocal Identities	Quotient Identities	Pythagorean Identities
$\csc \theta = \frac{1}{\sin \theta}$, $\sin \theta \neq 0$	$\tan \theta = \frac{\sin \theta}{\cos \theta}$, $\cos \theta \neq 0$	$\sin^2 \theta + \cos^2 \theta = 1$
$\sec \theta = \frac{1}{\cos \theta}$, $\cos \theta \neq 0$	$\cot \theta = \frac{\cos \theta}{\sin \theta}$, $\sin \theta \neq 0$	$1 + \tan^2 \theta = \sec^2 \theta$
$\cot \theta = \frac{1}{\tan \theta}$, $\tan \theta \neq 0$		$1 + \cot^2 \theta = \csc^2 \theta$

2. Suppose that in solving a problem you encounter two expressions, for example, $\sin^2 \theta$ and $(1 - \sin^2 \theta) \cdot \tan^2 \theta$, and you want to determine whether or not they are equivalent for all values θ for which both expressions are defined.

a. How could you use tables to support or contradict the claim that the equation $(1 - \sin^2 \theta) \cdot \tan^2 \theta = \sin^2 \theta$ is an identity?

b. How could you use graphs to support or contradict the claim that the equation is an identity?

c. Definitions and fundamental identities can be used to show that one expression can be transformed into the other. Justify each step in the following reasoning chain.

$$(1 - \sin^2 \theta) \cdot \tan^2 \theta = (1 - \sin^2 \theta) \cdot \frac{\sin^2 \theta}{\cos^2 \theta}$$
$$= \cos^2 \theta \cdot \frac{\sin^2 \theta}{\cos^2 \theta}$$
$$= \sin^2 \theta$$

d. Explain why the symbolic reasoning used above *proves* the equivalence of $\sin^2 \theta$ and $(1 - \sin^2 \theta) \cdot \tan^2 \theta$, while the approaches involving tables or graphs only make it plausible that the equation is an identity, but do not prove it.

Proving an identity involves a symbolic reasoning strategy in which each step can be justified by citing a definition or previously proved identity. The reasoning strategy used in Activity 2 Part c was to transform the left and more complicated side of $(1 - \sin^2 \theta) \cdot \tan^2 \theta = \sin^2 \theta$ into the simpler right side using known facts. Sometimes it is helpful to transform each side of a proposed identity independently into a third equivalent form, as illustrated in Activity 3.

3. Study the following proof of the identity $(1 - \sin\theta)(1 + \sin\theta) = \dfrac{1}{1 + \tan^2\theta}$, in which the expression on each side is transformed independently of the other. The vertical line is used to emphasize this fact.

$$
\begin{array}{c|c}
(1 - \sin\theta)(1 + \sin\theta) & \dfrac{1}{1 + \tan^2\theta} \\[2mm]
1 - \sin^2\theta & \dfrac{1}{\sec^2\theta} \\[2mm]
\cos^2\theta & \dfrac{1}{\sec\theta}\cdot\dfrac{1}{\sec\theta} \\[2mm]
 & \cos\theta\cdot\cos\theta \\[2mm]
 & \cos^2\theta
\end{array}
$$

Since each step on the right side can be reversed, it follows that
$$(1 - \sin\theta)(1 + \sin\theta) = \frac{1}{1 + \tan^2\theta}.$$

   a. Justify each step in the independent manipulations of each side.

   b. How does the fact that each step on the right side can be reversed complete the proof of the identity?

   c. Why would it be inappropriate to treat the proposed identity as an equation and then use associated properties of equality to prove it?

4. Now consider the identity $\tan\theta \cdot \sin\theta = \sec\theta - \cos\theta$.

   a. Prove this identity by showing each side is equivalent to $\dfrac{\sin^2\theta}{\cos\theta}$.

   b. Prove this identity by transforming the right side into the form on the left side.

   c. Look back at your two proofs. Which proof strategy do you prefer in this case and why?

5. Decide whether each of the following equations is or is not an identity. If an equation is an identity, use symbolic reasoning to prove it. If not, provide a counterexample.

   a. $\sec\theta - \tan\theta \sin\theta = \cos\theta$

   b. $\tan^2\theta - 2\sec\theta \sin\theta = \tan\theta\,(\tan\theta - 2)$

   c. $\cot\theta = \cos\theta \csc\theta$

   d. $\tan^2\theta = \dfrac{1 - \cos^2\theta}{\cos^2\theta}$

   e. $\csc^2\theta - \sec^2\theta = 1$

   f. $\sec^2\theta = \dfrac{\sin^2\theta + \cos^2\theta}{\cos^2\theta}$

   g. $1 - \tan\theta = \dfrac{\cos\theta - \sin\theta}{\cos\theta}$

   h. $(1 - \sin^2\theta)(1 + \tan^2\theta) = 1$

   i. $\cos(\theta_1 + \theta_2) = \cos\theta_1 + \cos\theta_2$

   j. $2\sin^2\theta - 1 = 1 - 2\cos^2\theta$

   k. $\sin\theta\,(\csc\theta - \sin\theta) = \cos^2\theta$

   l. $\sec^2\theta - \csc^2\theta = \dfrac{\tan\theta - \cot\theta}{\sin\theta \cos\theta}$

Trigonometric identities are special types of equations involving trigonometric functions.

ⓐ How is a trigonometric identity similar to, and different from, an algebraic property like the associative property for addition?

ⓑ How is a trigonometric identity different from a trigonometric equation?

ⓒ Explain why examining tables of values and/or graphs of functions is not sufficient to prove an equation is an identity.

ⓓ Describe how you would prove an equation is an identity.

Be prepared to share your comparisons, explanations, and descriptions with the entire class.

On Your Own

Prove each statement is or is not an identity.

a. $\dfrac{1}{\cos\theta\,\sin\theta} - \dfrac{\cos\theta}{\sin\theta} = \tan\theta$　　　**b.** $\sin\theta = \sqrt{1 - \cos^2\theta}$

c. $\csc\theta + \cot\theta = \dfrac{1 + \cos\theta}{\sin\theta}$　　　**d.** $\cos^2\theta - \sin^2\theta = \cos^2\theta - \dfrac{1}{2}$

INVESTIGATION 3 ▸ Sum and Difference Identities

In your previous work in mathematics, you have often described the location of points in a plane in terms of *rectangular coordinates* (x, y). In this case, x is the horizontal directed distance of the point from the y-axis and y is the vertical directed distance from the x-axis. Sometimes, as in the case of images on radar screens for example, it is more useful to specify the location of a point in terms of *polar coordinates* (r, θ). Here r is the length and θ is the angle of the position vector $\overrightarrow{OP}$, as in the diagram at the right.

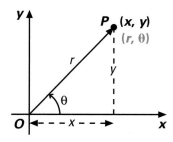

1. Referring to the diagram above, explain why the coordinates (x, y) of a point P in a rectangular coordinate system are related to the coordinates (r, θ) of the point in a polar coordinate system by the following equations:

$$x = r\cos\theta \text{ and } y = r\sin\theta, \text{ where } r = \sqrt{x^2 + y^2}$$

 a. Determine the rectangular coordinates of each point with the given polar coordinates: $A\left(10, \frac{\pi}{2}\right)$, $B(5, 120°)$, $C(4, 315°)$.

 b. Determine the polar coordinates of each point with the given rectangular coordinates: $P(3, 8)$, $Q(-5, 0)$, $R(-6, 8)$.

2. Now examine the diagram at the right in which:

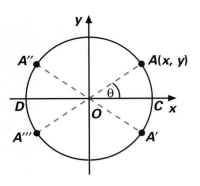

- Point A' is the image of point A reflected across the x-axis;

- Point A'' is the image of point A reflected across the y-axis;

- Point A''' is the image of point A rotated $180°$ about the origin.

θ is the measure of $\angle AOC$ and r is the length of $\overline{OA}$.

a. Determine the polar coordinates of the points A, A', A'', and A'''.

b. Explain the reasoning in the following derivation of the *opposite-angle identities*:

$$\cos(-\theta) = \cos\theta \text{ and } \sin(-\theta) = -\sin\theta$$

Since the rectangular coordinates of point A are $(r\cos\theta, r\sin\theta)$, the rectangular coordinates of point A' are $(r\cos\theta, -r\sin\theta)$. Since A has polar coordinates (r, θ), A' has polar coordinates $(r, -\theta)$ and thus A' has rectangular coordinates $(r\cos(-\theta), r\sin(-\theta))$. It follows that $r\cos(-\theta) = r\cos\theta$ and $\cos(-\theta) = \cos\theta$. Similarly, $\sin(-\theta) = -\sin\theta$.

c. Use the fact that $m\angle A''OC = \pi - \theta$ and reasoning similar to that above to derive identities for $\cos(\pi - \theta)$ and $\sin(\pi - \theta)$ in terms of $\cos\theta$ and $\sin\theta$. Explain in terms of function graphs why these identities make sense.

d. Find identities relating $\cos(\pi + \theta)$ and $\sin(\pi + \theta)$ to $\cos\theta$ and $\sin\theta$ respectively. Explain graphically why the identities you derived make sense.

e. Derive identities for $\tan(-\theta)$, $\tan(\pi - \theta)$, and $\tan(\pi + \theta)$. Interpret your identities graphically.

3. In Activity 2, you saw how line symmetry involving the coordinate axes or rotational symmetry about the origin are useful tools in deriving identities. In this activity, you will see how symmetry about the line $y = x$ is also useful.

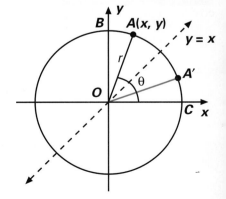

a. In the diagram at the right, $A(x, y) = A(r\cos\theta, r\sin\theta)$ and the polar coordinates of point A are $A(r, \theta)$. Reflect point A across the line $y = x$. Find the rectangular and polar coordinates of the image point A'.

b. Reasoning from your results in Part a, write identities relating the sine and cosine of $\frac{\pi}{2} - \theta$ to those of $\cos\theta$ and $\sin\theta$.

c. How do these identities differ from those found in Activity 2?

In your previous work, you used vectors to model navigation problems. For example, suppose two ships leave a harbor at approximately the same time and travel in different directions as shown in the diagram at the right. (The symbols "α" and "β" are the Greek letters "alpha" and "beta," respectively). To determine how far apart the ships are at any point in time, you could use the Law of Cosines. This use of the Law of Cosines would require that you find the value of $\cos(\alpha - \beta)$. In the next activity, you will examine a derivation of a useful identity involving $\cos(\alpha - \beta)$.

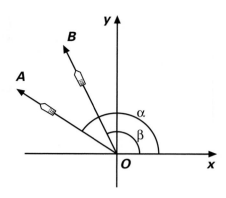

4. Suppose $\overrightarrow{OA}$ and $\overrightarrow{OB}$ in the diagram at the right make angles of α and β with the positive x-axis and that the lengths of $\overrightarrow{OA}$ and $\overrightarrow{OB}$ are r.

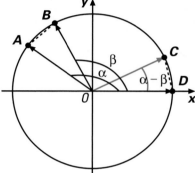

   a. Find the rectangular and polar coordinates of points A, B, C, and D on circle O where $\overrightarrow{OC}$ makes an angle of $\alpha - \beta$ with the positive x-axis and point D is on the x-axis.

   b. Analyze how the distance formula is used below to derive an identity involving $\cos(\alpha - \beta)$. Supply reasons for each step in the derivation.

   Since $AB = CD$, by the distance formula,

$$\sqrt{(r\cos\alpha - r\cos\beta)^2 + (r\sin\alpha - r\sin\beta)^2} = \sqrt{[r\cos(\alpha - \beta) - r]^2 + [r\sin(\alpha - \beta)]^2}.$$

   Now square both sides and simplify.

$$(r\cos\alpha - r\cos\beta)^2 + (r\sin\alpha - r\sin\beta)^2 = [r\cos(\alpha - \beta) - r]^2 + [r\sin(\alpha - \beta)]^2$$
$$(\cos\alpha - \cos\beta)^2 + (\sin\alpha - \sin\beta)^2 = [\cos(\alpha - \beta) - 1]^2 + \sin^2(\alpha - \beta)$$
$$\cos^2\alpha + \sin^2\alpha + \cos^2\beta + \sin^2\beta - 2(\cos\alpha\cos\beta + \sin\alpha\sin\beta) = \cos^2(\alpha - \beta) + \sin^2(\alpha - \beta) - 2\cos(\alpha - \beta) + 1$$
$$1 + 1 - 2(\cos\alpha\cos\beta + \sin\alpha\sin\beta) = 1 - 2\cos(\alpha - \beta) + 1$$
$$\cos\alpha\cos\beta + \sin\alpha\sin\beta = \cos(\alpha - \beta)$$

   c. How could you use graphs to illustrate the identity $\cos(\alpha - \beta) = \cos\alpha\cos\beta + \sin\alpha\sin\beta$?

5. The *difference identity* $\cos(\alpha - \beta) = \cos\alpha\cos\beta + \sin\alpha\sin\beta$ can be used to derive other useful identities.

   a. Use the fact that $\alpha + \beta = \alpha - (-\beta)$ to derive a *sum identity* for $\cos(\alpha + \beta)$. Begin as follows: $\cos(\alpha + \beta) = \cos[\alpha - (-\beta)]$.

b. Use the fact that $\cos\left(\frac{\pi}{2} - \theta\right) = \sin\theta$ to derive an identity for $\sin(\alpha + \beta)$. One possible way to begin is as follows:

$$\sin(\alpha + \beta) = \cos\left[\frac{\pi}{2} - (\alpha + \beta)\right] = \cos\left[\left(\frac{\pi}{2} - \alpha\right) - \beta\right]$$

c. Use the technique suggested in Part a to derive a difference identity for $\sin(\alpha - \beta)$.

d. Prove the following identities.

    i. $\cos(2\pi - \theta) = \cos\theta$ **ii.** $\sin(2\pi - \theta) = -\sin\theta$

e. Explain why the identities in Part d make sense geometrically in terms of a unit circle. In terms of function graphs.

6. Using relationships from geometry, you can find exact values of the trigonometric functions for angles whose measures are multiples of 30° or 45°, that is $\frac{\pi}{6}$ or $\frac{\pi}{4}$. By using sums and differences of these angle measures, you can find exact values of the functions for many other angles.

a. In the diagram below, $\triangle ABC$ is an equilateral triangle with side length 2 and $\overline{BD}$ is an altitude. Also, $\triangle PQR$ is an isosceles right triangle with legs of length 1. In each case, explain how the other indicated side lengths and angle measures can be determined.

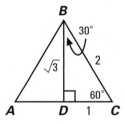
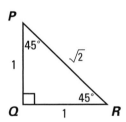

b. Use the diagrams above to help you complete a copy of the table below giving exact values of $\sin\theta$ and $\cos\theta$ for selected values of θ (in degrees and radians). *Note*: $\sin 45°$ is expressed as $\frac{\sqrt{2}}{2}$ since $\frac{1}{\sqrt{2}} = \frac{1}{\sqrt{2}} \cdot \frac{\sqrt{2}}{\sqrt{2}} = \frac{\sqrt{2}}{2}$ and the latter form is easier to interpret.

θ	0° or 0	30° or $\frac{\pi}{6}$	45° or $\frac{\pi}{4}$	60° or $\frac{\pi}{3}$	90° or $\frac{\pi}{2}$	180° or π	270° or $\frac{3\pi}{2}$
$\sin\theta$	?	?	$\frac{\sqrt{2}}{2}$	?	?	?	?
$\cos\theta$	?	?	?	?	?	?	?

c. Use your completed table and the identities from Activities 4 and 5 to find exact values for each expression below.

    i. $\cos 15°$ **ii.** $\cos\frac{7\pi}{12}$ **iii.** $\sin\frac{\pi}{12}$

    iv. $\sin 105°$ **v.** $\cos\left(\pi - \frac{\pi}{6}\right)$ **vi.** $\sin 285°$

7. Use identities and symbolic reasoning to transform each expression into an equivalent expression as indicated.

a. Transform $\frac{\cos \theta - \sin \theta}{\cos \theta}$ into an expression involving only the tangent function.

b. Transform $\sin 2x \cos 3x + \cos 2x \sin 3x$ into an expression involving only the sine function.

c. Transform $\sin 3x \cos 2x - \cos 3x \sin 2x$ into an expression involving only the sine function.

d. Transform $\cos 2x \cos x - \sin 2x \sin x$ into an expression involving only the cosine function.

e. Transform $2 \cos^2 x - \sin^2 x$ into an expression involving only:

    i. the sine function   **ii.** the cosine function

Checkpoint

In this investigation, you have added opposite angle identities and sum and difference identities for the sine and cosine functions to your toolkit of useful identities. Proofs of these identities involved moving flexibly between rectangular and polar coordinates of points.

ⓐ Write identities for $\sin (-\alpha)$ and $\cos (-\alpha)$.

ⓑ Write identities for $\sin (\alpha + \beta)$, $\cos (\alpha + \beta)$, $\sin (\alpha - \beta)$, and $\cos (\alpha - \beta)$.

ⓒ Suppose $P(x, y) = P(r, \theta)$. How do you calculate x and y when given r and θ? How do you calculate r and θ when given x and y?

Be prepared to compare your identities and methods to those of other groups.

▶ On Your Own

Reflect on the mathematical methods in this investigation as you complete the following tasks.

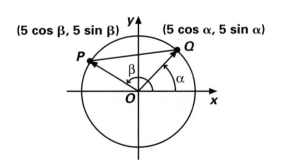

a. Using the diagram and information at the right:

    ■ What is the radius of the circle?

    ■ Find an expression for the length PQ.

    ■ What are the polar coordinates of points P and Q?

b. Use symbolic reasoning to derive an identity expressing $\sin \left(\frac{3\pi}{2} - \alpha \right)$ in terms of the sine or cosine of α. Repeat for $\cos \left(\frac{3\pi}{2} - \alpha \right)$.

c. Derive an identity expressing $\sin (90° + \alpha)$ in terms of the sine or cosine of α. Repeat for $\cos (90° + \alpha)$.

Modeling

1. Recall that the horizontal distance, in feet, that a projectile such as a golf ball or soccer ball will travel in the air is given by $R(\theta) = \frac{V^2 \cos \theta \sin \theta}{16}$, where V is given in feet per second.

   a. Using the fact that the horizontal and vertical components of a kicked ball are given by $d(t) = (V \cos \theta) \cdot t$ and $h(t) = (V \sin \theta) \cdot t - 16t^2$, respectively, prove that $R(\theta) = \frac{V^2 \cos \theta \sin \theta}{16}$.

   b. Prove the identity $\sin 2\theta = 2 \sin \theta \cos \theta$.

   c. Prove that the range formula above can be re-expressed in simpler form as $R(\theta) = \frac{V^2 \sin 2\theta}{32}$.

   d. Suppose a golfer can consistently hit the ball so that it leaves the tee with an initial velocity V. Prove that the maximum distance is attained when $\theta = 45°$.

   e. Imagine kicking a soccer ball with an initial velocity V. Prove that the horizontal distance the ball will travel in the air will be the same for $\theta = 45° + \alpha$ as for $\theta = 45° - \alpha$.

2. A two-phase alternating current circuit is supplied power by two generators. The voltages produced by the two generators are given by

$$v_1 = \sqrt{2}V_p \cos \left(\theta + \frac{\pi}{p} \right) \text{ and } v_2 = \sqrt{2}V_p \cos \left(\theta - \frac{\pi}{p} \right).$$

The average voltage, $\frac{v_1 + v_2}{2}$, is reported as $\frac{v_1 + v_2}{2} = \sqrt{2}V_p \cos \theta \cos \frac{\pi}{p}$. Prove that the reported average is correct.

3. A frequently used equation in the electromagnetic wave theory of light is

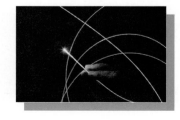

$$E'' = -E \cdot \left(\frac{k \cos r - \cos i}{k \cos r + \cos i} \right)$$

where k is the index of refraction, i is the angle of incidence, r is the angle of refraction, and $k = \frac{\sin i}{\sin r}$. Prove that the above equation is equivalent to

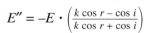

$$E'' = -E \cdot \left(\frac{\sin (i - r)}{\sin (i + r)} \right)$$

4. As indicated in Modeling Tasks 1–3, in applications involving trigonometric functions it is often necessary or useful to rewrite trigonometric expressions in equivalent forms, often using only a single trigonometric function. This task will provide additional practice with symbolic reasoning strategies that are helpful in such situations. For each of the following equations, prove or disprove that it is an identity.

    a. $\frac{\sec \theta + 1}{\sec \theta} = 1 + \cos \theta$

    b. $\frac{1}{1 + \cot^2 \theta} = (1 + \cos \theta)(1 - \cos \theta)$

    c. $\cos (\pi + \theta) = -\cos \theta$

    d. $\frac{\sec \theta \sin \theta}{\tan \theta + \cot \theta} = \sin^2 \theta$

    e. $\cos (\pi - \theta) = \cos (\pi + \theta)$

    f. $\cos^2 \theta - \sin^2 \theta = 1 - 2 \sin^2 \theta$

    g. $\sin (\pi - \theta) = -\sin (\pi + \theta)$

    h. $\frac{2 \tan \theta}{1 + \tan^2 \theta} = 2 \sin \theta \cos \theta$

5. Prove each identity.

    a. $\cos (\alpha + \beta) - \cos (\alpha - \beta) = -2 \sin \alpha \sin \beta$

    b. $\cos (\alpha + \beta) \cos (\alpha - \beta) = \cos^2 \alpha - \sin^2 \beta$

    c. $\sin (\alpha + \beta) + \sin (\alpha - \beta) = 2 \sin \alpha \sin \beta$

    d. $\sin (\alpha + \beta) \sin (\alpha - \beta) = \sin^2 \alpha - \sin^2 \beta$

Organizing

1. If $\csc \theta = \frac{13}{5}$ and $\frac{\pi}{2} < \theta < \pi$, evaluate the five remaining trigonometric functions of θ.

2. Prove the following sum and difference identities for the tangent function.

    a. $\tan (\alpha + \beta) = \frac{\tan \alpha + \tan \beta}{1 - \tan \alpha \tan \beta}$

    b. $\tan (\alpha - \beta) = \frac{\tan \alpha - \tan \beta}{1 + \tan \alpha \tan \beta}$

3. Derive identities in terms of trigonometric functions of α for each expression below. These identities are sometimes referred to as *double-angle identities*. You were asked to prove the double-angle identity for the sine function in Modeling Task 1.

    a. $\cos 2\alpha$

    b. $\tan 2\alpha$

4. Prove that $\cos 2\alpha$ is equivalent to each of the following expressions.

    a. $\cos^2 \alpha - \sin^2 \alpha$

    b. $2 \cos^2 \alpha - 1$

    c. $1 - 2 \sin^2 \alpha$

5. A function f is called an **even function** provided $f(-x) = f(x)$ for all values of x in the domain. A function g is called an **odd function** provided $g(-x) = -g(x)$ for all values of x in the domain.

   a. Explain why the cosine function is an even function.

   b. Explain why the sine function is an odd function.

   c. Is the tangent function an even function, an odd function, or neither? Justify your answer.

   d. Is the absolute value function $a(x) = |x|$ an even function, an odd function, or neither? Explain.

   e. What is true about the graph of any even function? Of any odd function? Why does this make sense?

   f. Give an example of a polynomial function that is an odd function. A polynomial function that is an even function. A polynomial function that is neither.

   g. Formulate and justify a generalization about polynomial functions that are odd functions and those that are even functions.

6. Prove each statement is or is not an identity.

   a. $\tan \alpha = \dfrac{\sin 2\alpha}{1 + \cos 2\alpha}$

   b. $\cos 2\alpha = \dfrac{1 - \tan^2 \alpha}{1 + \tan^2 \alpha}$

   c. $\tan \alpha = \cot 2\alpha - \csc 2\alpha$

   d. $\sin 2\alpha = \dfrac{2 \tan \alpha}{1 + \tan^2 \alpha}$

Reflecting

1. Explain why an identity can be thought of as a special type of statement of equality.

2. What do you need to do to prove that an equation is *not* an identity? To prove that an equation is an identity?

3. Sometimes it is suggested to students that a good strategy for proving a trigonometric identity is to first change all functions to sines and cosines. Is this always possible? Why might it seem like a useful idea? A bad idea?

4. Why must proof of an identity be based on symbolic reasoning rather than reasoning from graphs or tables? What role might tables and graphs play in exploring identities?

5. Approximations involving trigonometric functions are often used in applied fields that involve radian measures of small angles. Assume that angles can be measured accurately to two decimal places.

a. The approximation sin θ ≈ θ is used in the field of optics. If approximations must be correct to two decimal places, find the largest acute angle for which the relationship holds.

b. The approximation tan θ ≈ θ is used in the field of mechanics. If approximations must be correct to two decimal places, find the largest acute angle for which this relationship holds.

c. Why are the relationships in Parts a and b not identities?

6. In Investigation 3, you proved that cos (π + θ) = −cos θ. But you might have asked yourself what clues may have suggested that this was an identity?

a. Explain why this identity is suggested by the unit circle interpretation of the cosine function.

b. Now think about the relationship in terms of function graphs.

■ How are the graphs of y = cos (π + θ) and y = cos θ related?

■ Why is it reasonable that the graph of y = cos (π + θ) is the same as the graph of y = −cos θ?

c. Use graphical reasoning to help you write an identity statement for csc (π + θ).

■ Does your identity make sense in terms of the unit circle interpretation of the trigonometric function?

■ Prove your identity.

Extending

1. In mathematics, there is often more than one way to prove a statement is true. Yihnan David Gau proposed the following visual proof of the identity sin 2t = 2 sin t cos t. (Source: "Proof Without Words: Double Angle Formulas," *Mathematics Magazine*, Vol. 71, no. 5 (December 1998), p. 385.) Explain how the diagrams below prove the identity. For what values of *t* is this proof valid?

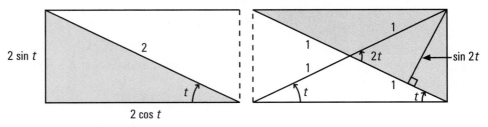

2 sin t cos t   sin 2t

2. When a point $A(x, y)$ is rotated by an angle θ about the origin to the point $A'(x', y')$, the new coordinates are $x' = x \cos \theta - y \sin \theta$ and $y' = x \sin \theta + y \cos \theta$. This can be represented in matrix form as follows:

$$\begin{bmatrix} x' \\ y' \end{bmatrix} = \begin{bmatrix} \cos \theta & -\sin \theta \\ \sin \theta & \cos \theta \end{bmatrix} \begin{bmatrix} x \\ y \end{bmatrix} = R_\theta \begin{bmatrix} x \\ y \end{bmatrix}$$

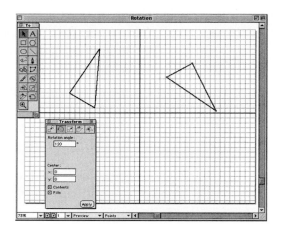

The matrix R_θ is a *rotation matrix*. The rotation of θ followed by a rotation of ϕ can be represented in matrix form by the product $R_\phi \cdot R_\theta$.

a. What is the magnitude of a rotation of θ followed by a rotation of ϕ?

b. According to the definition above, what is the matrix $R_{\phi + \theta}$?

c. $R_{\phi + \theta} = R_\phi \cdot R_\theta$. Carry out this multiplication.

d. How can Parts b and c be used to derive identities for $\sin (\phi + \theta)$ and for $\cos (\phi + \theta)$?

e. How could matrices be used to derive identities for $\sin (\phi - \theta)$ and for $\cos (\phi - \theta)$?

3. Prove each statement is or is not an identity.

a. $\csc 2\alpha = \dfrac{\sec \alpha \csc \alpha}{2}$

b. $\sec 2\alpha = \dfrac{\sec^2 \alpha}{2 - \sec^2 \alpha}$

c. $\cot 2\alpha = \dfrac{1 + \cos 4\alpha}{\sin 4\alpha}$

d. $\sin 3\alpha = 3 \sin \alpha - 4 \sin^3 \alpha$

e. $\dfrac{\sin 2\alpha}{\sin \alpha} - \dfrac{\cos 2\alpha}{\cos \alpha} = \sec \alpha$

f. $\sec^4 \theta - \tan^4 \theta = \sec^2 \theta + \tan^2 \theta$

g. $\cos 4\alpha = 8 \cos^4 \alpha - 8 \cos^2 \alpha + 1$

h. $\sin 4\alpha = 4 \sin \alpha \cos \alpha (2 \cos^2 \alpha - 1)$

4. A tree casts a shadow of 100 m at one time and 32 m at a later time when the angle of the line of sight to the sun and the horizontal is twice as large. Find the height of the tree.

5. Use the double-angle formulas (see Organizing Task 3) to derive the following *half-angle identities.* The ± sign indicates that the values are sometimes positive and sometimes negative, depending on the size of α.

  a. $\sin \frac{\alpha}{2} = \pm \sqrt{\frac{1 - \cos \alpha}{2}}$

  b. $\cos \frac{\alpha}{2} = \pm \sqrt{\frac{1 + \cos \alpha}{2}}$

  c. How would you decide whether to use the positive or negative value of the expression?

6. Recall that the average rate of change of $f(x)$ over the interval from x to $x + h$ is given by the difference quotient $\frac{f(x + h) - f(x)}{h}$. If the difference quotient approaches a limiting value as h gets small, then you obtain the instantaneous rate of change or the derivative of $f(x)$.

  a. Write an expression for the average rate of change of $f(x) = \cos x$ over the interval from x to $x + h$.

  b. Explore the graph of the average rate of change function when $h = 0.001$. What basic function has a similar graph?

  c. Use identities to write the difference quotient $\frac{\cos (x + h) - \cos x}{h}$ in the form $\frac{(\cos x)g(h) + (\sin x)q(h)}{h}$.

   ■ Investigate $\frac{g(h)}{h}$ and $\frac{q(h)}{h}$ to determine their limiting values as h gets very small.

   ■ What is the instantaneous rate of change of $f(x) = \cos x$?

  d. Repeat Parts a–c for $f(x) = \sin x$.

1. If $\frac{6}{x-9} = \frac{1}{4}$, then x is

   (a) 15   **(b)** 19   **(c)** -15 **(d)** 13   **(e)** 33

2. Multiply: $(x^2 + 1)(x - 1)$

   (a) $x^3 - 1$ **(b)** $x^3 - x^2 + x - 1$ **(c)** $x^2 + x - 1$

   (d) $2x^2 - x^2 + x - 1$ **(e)** $x^3 + 1$

3. What is the slope of the line shown below?

   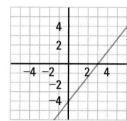

   (a) $\frac{3}{4}$ **(b)** $-\frac{3}{4}$

   (c) $-\frac{4}{3}$ **(d)** $\frac{4}{3}$

   (e) $-\frac{3}{5}$

4. If $f(x) = 3x^4 + 2x^2 - 5x + 6$, then $f(7)$ is

   (a) 3,936   **(b)** 7,272   **(c)** 7,342   **(d)** $-7,342$ **(e)** $-7,272$

5. Solve: $2(x + 9)^2 + 17(x + 9) + 35 = 0$

   (a) $x = 4$ or $x = 6.5$ **(b)** $x = -3.5$ or $x = -5$ **(c)** $x = 3.5$ or $x = 5$

   (d) $x = 14$ or $x = 12.5$ **(e)** $x = -14$ or $x = -12.5$

6. This is the graph of which inequality?

   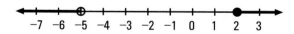

   (a) $-5 \le x \le 2$ **(b)** $x > -5$ or $x \le 2$ **(c)** $x < -5$ or $x \ge 2$

   (d) $x \le -5$ or $x > 2$ **(e)** $x \le -5$ or $x \ge 2$

7. If $f(x) = \frac{(x^2 + 3)(x + 2)}{x + 1}$, then the graph of the function

   (a) has a vertical asymptote at $x = 1$.

   (b) has a horizontal asymptote at $x = -1$.

   (c) is a parabola.

   (d) has a vertical asymptote at $x = -3$.

   (e) crosses the x-axis exactly one time.

8. $\tan 3x =$

   (a) $\tan 2x + \tan x$ (b) $\frac{\sin 3x}{\cos 3x}$ (c) $\frac{\cos 3x}{\sin 3x}$

   (d) $\sec 3x - 1$ (e) $\tan (90° - 3x)$

9. If $\log x = \log y + \log w$, then $w =$

   (a) $\frac{x}{y}$ (b) xy (c) $x - y$ (d) $x + y$ (e) $\frac{y}{x}$

10. Which of the following could be a portion of the graph of $y = 3 + \sqrt{x + 5}$?

   (a) 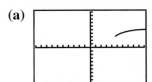   (b) 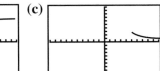   (c)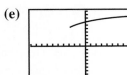

   (d) 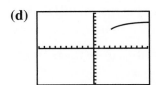   (e)

Solving Trigonometric Equations

Solving equations is an integral part of using functions to analyze problem situations. In your previous coursework in mathematics, you saw that a variation of the sine (or cosine) function can be used to model the cyclical pattern in temperatures for a given locale. For example, the mean daily Fahrenheit temperature in Fairbanks, Alaska can be modeled by the function $T(x) = 37 \sin\left(\frac{2\pi}{365}(x - 101)\right) + 25$, where x is the day of the year, with $x = 0$ representing January 1.

Fairbanks, Alaska

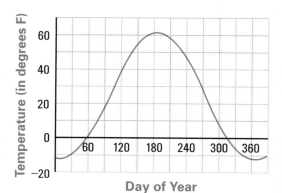

Think About This Situation

Writing and solving equations that match questions and then interpreting your solutions are important components of mathematical modeling.

a On what days would you predict the average temperature in Fairbanks to be 0°F? What equation matches this question?

b What symbolic reasoning strategies might you use to solve the equation in Part a?

c Think about how solving trigonometric equations using symbolic reasoning might be similar to solving linear equations.

- How would you solve $ax + b = c$ for x? How might you solve $a \sin \theta + b = c$ for θ?

- How many solutions does $ax + b = c$ have? How many solutions does $a \sin \theta + b = c$ have?

- Does $ax + b = c$ always have a solution? Does $a \sin \theta + b = c$ always have a solution?

As is the case with other kinds of equations, you can estimate the solutions of trigonometric equations by using tables or graphs. In this lesson, however, you will learn how to use symbolic reasoning strategies to find solutions. These strategies are similar to those used to solve linear and quadratic equations. However, they will also draw upon your knowledge of the periodicity of trigonometric functions, and, in some cases, your knowledge of basic trigonometric identities.

INVESTIGATION 1 Solving Linear Trigonometric Equations

In this investigation, you will solve equations like $af(x) = b$, where $f(x)$ is a trigonometric function. Given $af(x) = b$, an initial question to ask is whether solutions even exist. If solutions exist, then how do you find them?

1. The question of whether a trigonometric equation has one or more solutions can often be answered by examining the symbolic form and using your understanding of the basic trigonometric functions. For each equation below, first indicate whether or not a solution exists and then explain your reasoning. You do not need to solve these equations.

   a. $3 \sin x = -1$ **b.** $3 \sin x = -4$

   c. $3 \tan x = -1$ **d.** $3 \tan x = -4$

   e. $3 \sec x = -1$ **f.** $3 \sec x = -4$

2. Suppose you are given a trigonometric equation in the form $af(x) = b$. What conditions on a and b ensure that $af(x) = b$ has a solution in each of the following cases?

   a. $f(x) = \sin x$ **b.** $f(x) = \sec x$ **c.** $f(x) = \tan x$

3. Now consider solving the equation $2 \sin x + 3 = 4$.

   a. Does this equation have a solution? Explain your reasoning.

   b. Use symbolic reasoning and the inverse trigonometric function capability of your calculator to find one solution of $2 \sin x + 3 = 4$ on the interval $0 \le x < 2\pi$. What is the solution in degrees?

   c. Use your table of exact trigonometric function values from Investigation 3 of Lesson 2 (page 470) to find an exact radian solution.

   d. Are there additional solutions on the interval $0 \le x < 2\pi$? Find all such solutions and record them in approximate radian, exact radian, and degree forms. Explain how you found each solution.

   e. Students in a class at Andover High School reported insightful ways of finding a second solution to $2 \sin x + 3 = 4$ on the interval $0 \le x < 2\pi$. For example, Jacob reasoned as follows: I know the graph of $y = \sin x$ is symmetric to the line $x = \frac{\pi}{2}$. Since $\frac{\pi}{6}$ is a solution of $2 \sin x + 3 = 4$, another solution is $\frac{\pi}{3}$ units to the right of $\frac{\pi}{2}$, namely $\frac{\pi}{2} + \frac{\pi}{3} = \frac{5\pi}{6}$.

   - Is the graph of $y = \sin x$ symmetric about the line $x = \frac{\pi}{2}$?

   - Why is the second solution $\frac{\pi}{3}$ radians to the right of $\frac{\pi}{2}$?

Sheila, a student in the same class, reported another way to find the second solution by using an identity. She noted that since $\sin(\pi - x) = \sin x$ and $x = \frac{\pi}{6}$ is a solution, $\pi - \frac{\pi}{6} = \frac{5\pi}{6}$ is also a solution.

- Does $\sin(\pi - x) = \sin x$? Explain.

- Does Sheila's argument prove $\frac{5\pi}{6}$ is a solution? Why or why not?

- Compare Jacob's and Sheila's reasoning strategies.

f. Suppose you wished to solve $2 \sin x + 3 = 4$ over the entire domain of $f(x) = \sin x$.

- Explain why $\frac{\pi}{6} + 2\pi n$, for any integer n, describes some of the solutions.

- Write an expression representing the remaining solutions.

4. Consider a similar equation involving the tangent function: $2 \tan x + 4 = 3$.

   a. Find one solution expressed in degrees.

   b. Write an expression giving all the possible solutions. How does the period of the tangent function help in writing this expression?

In the previous two activities, you found all the solutions to two trigonometric equations. For the equation $2 \tan x + 4 = 3$, you found one solution, for example $153.4°$ or $-26.6°$, and wrote an expression giving all solutions by adding and subtracting whole number multiples of the period, $180°$. The equation $2 \sin x + 3 = 4$ has two solutions, $\frac{\pi}{6}$ and $\frac{5\pi}{6}$, not related to each other by period 2π. You then used each of these and the fact that the period of the sine function is 2π to write two expressions that together described all solutions.

The key to solving a trigonometric equation of the form $f(x) = c$ is to find one or two solutions. If a solution exists, one can always be found by using the appropriate inverse function on your calculator. You can then use that solution to generate expressions for all solutions. Graphs can help guide your reasoning. The three inverse trigonometric functions on your calculator will produce solutions in the following ranges. In Reflecting Task 1 (page 489), you will examine reasons for these restricted ranges.

Ranges for Inverse Trigonometric Functions

$$-90° \leq \sin^{-1} x \leq 90° \quad \text{or} \quad -\frac{\pi}{2} \leq \sin^{-1} x \leq \frac{\pi}{2}$$

$$0° \leq \cos^{-1} x \leq 180° \quad \text{or} \quad 0 \leq \cos^{-1} x \leq \pi$$

$$-90° < \tan^{-1} x < 90° \quad \text{or} \quad -\frac{\pi}{2} < \tan^{-1} x < \frac{\pi}{2}$$

5. Now, consider the equation $2 \cos x + \sqrt{3} = 0$.

   a. Solve this equation for $\cos x$. Does the equation have a solution for x?

   b. Find two solutions in radians that are not related to each other by period 2π.

   c. Use your two solutions to write expressions giving all solutions.

6. Solve each trigonometric equation for all values of x. When possible, give exact solutions. Express your solutions in both radians and in degrees.

    a. $3 \sin x - 2 = 1$ **b.** $5 \cos x + 1 = 3$

    c. $2 \tan x - 3 = 2$ **d.** $3 \sec x = 2$

Note: For the remainder of this unit, express solutions to trigonometric equations in radians unless the context or given equation suggests otherwise.

7. You now have symbolic reasoning strategies to solve equations of the form $af(x) = b$, where $f(x)$ is a trigonometric function. Next consider strategies for solving $af(cx + d) = b$, where $f(x)$ is a trigonometric function. In particular, consider the equation $\sin 2x = \frac{1}{2}$.

    a. Explain why this equation is of the form $af(cx + d) = b$.

    b. Working together as a group, brainstorm a symbolic reasoning strategy to solve $\sin 2x = \frac{1}{2}$ for x. Compare your solution strategy with that of other groups.

    c. Compare your solution strategy to the following method which uses the idea of a *substitution of variable*.

        To solve $\sin 2x = \frac{1}{2}$, let $\theta = 2x$.

        Then, solve $\sin \theta = \frac{1}{2}$.

        From Activity 3, you know the solutions are

        $\theta = \frac{\pi}{6} + 2\pi n$ and $\theta = \frac{5\pi}{6} + 2\pi n$ for any integer n.

        To find x, replace θ with $2x$ and solve for x.

        So, $2x = \frac{\pi}{6} + 2\pi n$, and thus $x = \frac{\pi}{12} + \pi n$

        or $2x = \frac{5\pi}{6} + 2\pi n$, and thus $x = \frac{5\pi}{12} + \pi n$.

8. Use the substitution of variable technique to solve $\tan \left(2x + \frac{\pi}{2}\right) = 1$. Check your solutions graphically and numerically.

9. Use symbolic reasoning strategies of your choice to solve each of the following trigonometric equations.

    a. $2 \cos (x + 30°) = 1$ **b.** $2 \cos (2x + 30°) = 1$

    c. $\sqrt{3} \tan 3x = 1$ **d.** $2 \sin \left(2x + \frac{\pi}{6}\right) = \sqrt{3}$

    e. $3 \sin (0.5x + 1) = 2$

10. In the "Think About This Situation" for this lesson, the mean daily Fahrenheit temperature in Fairbanks, Alaska was modeled by the following function:

$$T(x) = 37 \sin \left[\frac{2\pi}{365} (x - 101)\right] + 25$$

    a. Use symbolic reasoning to predict on what days of the year the mean temperature in Fairbanks will be 0°F. How will you decide whether to set your calculator in degree or radian mode?

    b. Use symbolic reasoning to predict on what days of the year the mean temperature in Fairbanks will be 40°F or above.

In this investigation, you have examined strategies for solving linear trigonometric equations.

a How is solving the equation $a \tan x + b = c$, which is linear in $\tan x$, similar to, and different from, solving $ax + b = c$, which is linear in x?

b Describe the procedures you would use to solve linear trigonometric equations such as the following:

- $a \cos (x - b) = c$
- $a \tan (x - b) = c$

c Explain why equations linear in a trigonometric function may have no solution or an infinite number of solutions.

Be prepared to share your thinking and procedures with the class.

On Your Own

Because of ocean tides, the depth of the River Thames in London varies sinusoidally as a function of time. Suppose the depth in meters as a function of the hour of the day is modeled by

$$d(t) = 3 \sin \left(\frac{\pi}{6}(t - 4) \right) + 8,$$

where $t = 0$ corresponds to 12:00 A.M. Use symbolic reasoning to complete the following tasks. Check your solutions graphically and numerically.

The River Thames in London.

a. Predict the minimum depth of the river and the times at which it occurs.

b. Predict the depth of the river at 2:00 P.M.

c. At approximately what times is the depth about 10 m?

d. During what time periods will the depth of the river exceed 8 m?

INVESTIGATION 2 ▶ Using Identities to Solve Trigonometric Equations

In the first investigation, you solved linear trigonometric equations involving sines, cosines, or tangents. In the following activities, you will investigate how to solve more complex trigonometric equations. The first step in solving these equations is to write a simpler, equivalent form involving a single function such as sine, cosine, or tangent.

1. Examine the following solution of $\sin 2x \cos x + \cos 2x \sin x = 1$.

   a. Give reasons for each step.

   $$\sin 2x \cos x + \cos 2x \sin x = 1$$
   $$\sin (2x + x) = 1$$
   $$\sin 3x = 1$$

   Let $3x = \theta$; then $\sin \theta = 1$.

   So, $\theta = \frac{\pi}{2} + 2\pi n$, for any integer n.

   Hence, $3x = \frac{\pi}{2} + 2\pi n$.

   Thus, $x = \frac{\pi}{6} + \frac{2\pi n}{3}$.

   b. Check the solution in the original equation for $n = 1, 2, -1$, and -2.

2. Give reasons for each step in the following solution of $\cos^2 x + \sin x + 1 = 0$.

   $$\cos^2 x + \sin x + 1 = 0$$
   $$1 - \sin^2 x + \sin x + 1 = 0$$
   $$\sin^2 x - \sin x - 2 = 0$$
   $$(\sin x - 2)(\sin x + 1) = 0$$

   So, $\sin x - 2 = 0$ or $\sin x + 1 = 0$

   $\sin x = 2$ has no real number solutions.

   If $\sin x + 1 = 0$, then $\sin x = -1$.

   Therefore, $x = \frac{3\pi}{2} + 2\pi n$, for any integer n.

3. Solve each of the following equations. Check your solutions.

   a. $2 \sin (x + 37°) = 1$

   b. $4 \cos^2 x = 1$

   c. $\cos 2x + \cos x = 0$

   d. $2 \sin x \cos x = \sqrt{2} \cos x$

   e. $2 \cos^2 x - 5 \cos x + 2 = 0$

   f. $\tan^2 x - \sec x - 1 = 0$

   g. $4 \sin x \cos x = \sqrt{3}$

   h. $\sin 2x \cos x - \cos 2x \sin x = -\frac{\sqrt{3}}{2}$

Solving trigonometric equations can be guided by your previous experiences in solving linear and quadratic polynomial equations. However, special consideration must be given to the special characteristics and properties of trigonometric functions.

a Describe the steps you would take, in the order you would do them, when solving the trigonometric equation $\sin 2x = \sin x$ for $0 \leq x < 2\pi$.

b How could you check your solutions in Part a?

c How would you modify your solutions in Part a if there were no restrictions on x?

Be prepared to explain your equation-solving method to the class.

On Your Own

Solve each of the following trigonometric equations for t. Check your solutions.

a. $6 \cos^2 t + 5 \cos t + 1 = 0$

b. $2 \cos^2 t = 3 \sin t + 3$

c. $2 \sin^2 3t - 5 \sin 3t = 3$

MORE
Modeling • Organizing • Reflecting • Extending

Modeling

Use symbolic reasoning in completing these Modeling tasks. Then make notes on how you could also answer the questions using graphs or tables.

1. After a pizza oven is heated to a specific temperature, for example, 500°F, the actual temperature in the oven oscillates from slightly below 500°F to slightly above 500°F. Suppose the temperature in one pizza oven is a function of time x in minutes with rule $t(x) = 10 \sin x + 500$.

   a. Why is this modeling function reasonable in terms of the context?

   b. When does the temperature in the oven first reach 505°F?

   c. At what other times is the temperature in the oven 505°F?

   d. How many times during the first hour will the temperature of the oven be 505°F?

2. Suppose a weight is suspended from the ceiling by a spring. When the weight is pulled down and released, it will oscillate up and down about its at-rest position. If the effects of friction and air resistance were ignored, the oscillations would repeat indefinitely. Suppose you pull the weight down 8 cm from its equilibrium position and let it go. The displacement of the weight from its at-rest position is a function of the number of seconds since it was released with rule $d(t) = -8 \cos 2\pi t$.

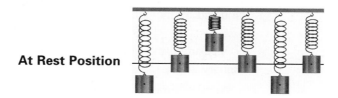

At Rest Position

a. How long will it take for the weight to make one complete oscillation?

b. When will the weight first be back at the same height from which it was released? Show two different ways to get this answer.

c. When will the weight be 3 cm below its at-rest position?

d. Will the weight ever be 10 cm below its at-rest position? If so, when? If not, why not?

3. The electricity used in most households of the United States is in the form of alternating currents. The voltage alternates smoothly between +110 and −110 volts, approximately 60 times per second. The modeling equation for this voltage is $V(t) = 110 \cos (120\pi t)$, where t is measured in seconds.

a. Explain why the coefficient of t is 120π.

b. At what values of t will the voltage be −50?

c. At what values of t will the voltage be 100?

4. As you have previously seen, the cyclic pattern of the water depth in a harbor can be modeled by a trigonometric function. Suppose the depth, in feet, in one harbor is given by $D(t) = 9 \cos \left(\frac{\pi}{6}t\right) + 15$, where t is measured in hours after the first high tide on a certain day.

a. Explain the meaning of 9, $\frac{\pi}{6}$, and 15 in the given equation.

b. At what times is the depth of the water 15 ft?

c. At what times is the depth of the water 8 ft?

d. At what times is the depth of the water 4 ft?

e. For how many hours each day will the water be deeper than 10 ft?

A harbor in Yorkshire, England.

5. As you saw in Unit 1, "Rates of Change," the populations of some animals vary with the time of year and can be modeled by trigonometric functions. Suppose that the rabbit population in Sleeping Bear Dunes National Forest, in thousands, is modeled by the equation $f(t) = 4.25 \sin\left(\frac{\pi}{6}t - \frac{\pi}{2}\right) + 8.25$, where t is measured in months since January 2000.

Sleeping Bear Dunes National Forest

a. At what times will the rabbit population be 6,000?

b. For what time periods during the year 2000 was the rabbit population less than 10,000?

Organizing

1. Solve each trigonometric equation.

a. $\sin x = -\frac{\sqrt{2}}{2}$

b. $\cos 2x = -\frac{\sqrt{2}}{2}$

c. $\sin\left(x + \frac{\pi}{4}\right) = -\frac{\sqrt{2}}{2}$

d. $\tan 0.5x = \sqrt{3}$

e. $\tan(x - 30°) = \sqrt{3}$

f. $2\sin x + \sqrt{3} = 0$

g. $4\tan x - 2 = 2$

h. $\sqrt{3}\sec x + 2 = 0$

2. Solve each trigonometric equation.

a. $4\sin^2 x - 3 = 0$

b. $\cos x + 2 = 3\cos x$

c. $2\sin^2 2x + \sin 2x = 0$

d. $\tan^2 3x + \tan 3x = 0$

e. $\sin^2 2x + 5\sin 2x + 6 = 0$

f. $\cos 2x - \sin x = 0$

g. $\sin 2x \sin x + \cos 2x \cos x = 1$

h. $\cos 2x + 3\cos x = 1$

3. If possible, find one exact zero for each of the following functions. Then look back at your work and describe how the zeroes for each function are related to the zeroes of $s(x) = \sin x$.

a. $f(x) = 2\sin x$

b. $g(x) = \sin x + 2$

c. $h(x) = \sin(x + 2)$

d. $j(x) = \sin 2x$

4. The solutions in degrees of $\sin x = \frac{1}{2}$ are $30° + 360°n$ and $150° + 360°n$, for any integer n.

   a. Find the solutions of $\sin 2x = \frac{1}{2}$.

   b. Find the solutions of $\sin 3x = \frac{1}{2}$.

   c. Find the solutions of $\sin 0.5x = \frac{1}{2}$.

   d. Describe how the solutions of $\sin ax = \frac{1}{2}$ are related to the solutions of $\sin x = \frac{1}{2}$.

5. The solutions in radians of $\tan x = 1$ are $\frac{\pi}{4} + \pi n$, for any integer n.

   a. Find the solutions of $\tan \left(x - \frac{\pi}{3}\right) = 1$.

   b. Find the solutions of $\tan \left(x - \frac{\pi}{2}\right) = 1$.

   c. Find the solutions of $\tan (x + 2) = 1$.

   d. Describe how the solutions of $\tan (x + a) = 1$ are related to the solutions of $\tan x = 1$.

Reflecting

1. Explain why the functions of $y = \sin x$, $y = \cos x$, and $y = \tan x$ do not have inverses.

   a. What restrictions on the domains of $y = \sin x$, $y = \cos x$, and $y = \tan x$ produce inverse functions?

   b. Explain how the restricted domains in Part a produce the ranges of $y = \sin^{-1} x$, $y = \cos^{-1} x$, and $y = \tan^{-1} x$ displayed on page 482.

2. What aspect of solving a linear trigonometric equation is the most difficult for you? How can or did you overcome that difficulty?

3. Esperanza solved the equation $\sin x = \cos x$ as follows:

   $$\sin x = \cos x$$
   $$\sin^2 x = \cos^2 x \text{ (squaring both sides)}$$
   $$\sin^2 x = 1 - \sin^2 x \text{ (identity)}$$
   $$2 \sin^2 x = 1$$
   $$\sin^2 x = \frac{1}{2} \text{ and } \sin x = \pm\frac{1}{\sqrt{2}}$$

   Thus, $x = 45° + 180°n$ and $x = 135° + 180°n$, for any integer n.

   a. Is Esperanza's reasoning correct? Explain.

   b. Is Esperanza's solution correct? If not, how can you guard against making similar errors?

4. Professional users of mathematics often use a Computer Algebra System (CAS) in their work. Once available only on computer, a CAS is now built into "super" calculators like the TI-92. A CAS allows the user to symbolically manipulate expressions and solve equations. Symbol sense and symbolic reasoning remain important in a CAS environment.

a. Examine the CAS display below. The marker ▶ on the right of the display in the second and third lines indicates that more follows. Do the solutions seem reasonable? Why or why not?

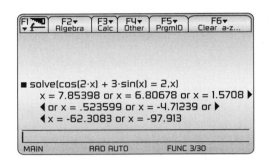

b. Describe how you would solve the general trigonometric equation $a \sin x + b = c$ for x.

■ How does your thinking allow you to assess the reasonableness of the CAS-produced result below?

■ Why do the solutions have the restriction $\left| \dfrac{b-c}{a} \right| \le 1$?

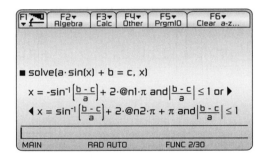

Extending

1. Illustrate two methods of solving the equation $2 \cos^2 x - \cos x - 1 = 0$, one using factoring directly and the other using substitution of variable ($u = \cos x$).

2. Solve each equation below using symbolic reasoning. Check your solutions by solving the equation using a different method.

a. $\cos^2 0.5x - 0.5 \cos x = 0.5$

b. $2 \sin^2 x + \sin x = 1$

    c. $4 \sin^4 x + \sin^2 x = 3$

    d. $\sin 2x + \cos 3x = 0$

3. In certain applications of calculus, it is necessary to express a product of two trigonometric functions as a sum or difference. This can often be accomplished by using one of the following *product-sum identities*. Use sum and difference identities for sine and cosine to prove these identities.

    a. $\cos \alpha \cos \beta = \frac{1}{2}[\cos (\alpha + \beta) + \cos (\alpha - \beta)]$

    b. $\sin \alpha \sin \beta = \frac{1}{2}[\cos (\alpha - \beta) - \cos (\alpha + \beta)]$

    c. $\sin \alpha \cos \beta = \frac{1}{2}[\sin (\alpha + \beta) + \sin (\alpha - \beta)]$

    d. $\cos \alpha \sin \beta = \frac{1}{2}[\sin (\alpha + \beta) - \sin (\alpha - \beta)]$

4. Some applications of calculus require that the sum or difference of two trigonometric functions be rewritten as a product. This can be done by using alternate forms of the product-sum identities. Derive these forms using the identities in Extending Task 3 and the substitution $\alpha = \frac{1}{2}(u + v)$ and $\beta = \frac{1}{2}(u - v)$. The four new identities relate the sum (difference) of sines and cosines of u and v to the product of sines and cosines of half sums and half differences. The identities are called *sum-product identities*.

5. Use the identities in Extending Tasks 3 and 4 as needed to solve each equation.

    a. $\sin x + \sin 2x + \sin 3x = 0$

    b. $\cos x - \cos 3x - \cos 5x = 0$

    c. $\sin x + \sin 2x - \sin 4x = 0$

1. What percent markup is there if a dress costing $40 sells for $62?

   (a) 30%   (b) 55%   (c) 65%   (d) 40%   (e) 50%

2. If $y = 1$, then $3\left|y\right| - \left|y - 4\right|$ equals

   (a) –2   (b) 2   (c) –6   (d) 8   (e) 0

3. The equation of a line parallel to the graph of $y = \frac{5}{4}x + 4$ is

   (a) $y = \frac{4}{5}x + 4$ (b) $y = -\frac{5}{4}x + 4$ (c) $y = -\frac{4}{5}x - 4$

   (d) $y = -\frac{5}{4}x - 4$ (e) none of these

4. Which of the following is a function?

   (a) $\{(5, -7), (-7, -6), (5, -8)\}$ (b) $\{(3, -8), (-4, -9), (-4, -6)\}$

   (c) $\{(5, -8), (-7, 3), (-7, -9)\}$ (d) $\{-4, -5, -6, -7\}$

   (e) $\{(-2, -1), (-1, -2), (-3, -2)\}$

5. If $f(x) = -(x - a)^2 + b$, then which of the following must be true?

   (a) The vertex is the maximum point and is at $(-a, b)$.

   (b) The vertex is the minimum point and is at $(-a, b)$.

   (c) The vertex is below the x-axis.

   (d) The vertex is the maximum point and is at (a, b).

   (e) The vertex is the minimum point and is at (a, b).

6. If $A = P(1 + r)^t$, then $P =$

(a) $\dfrac{(1 + r)^t}{A}$ **(b)** $\dfrac{A}{(1 + r)^t}$ **(c)** $(1 + r)^t - A$

(d) $A - (1 + r)^t$ **(e)** $A(1 + r)^t$

7. The polynomial equation $x^4(x^2 - 16)(x^2 - 10) = 0$ has how many distinct real roots?

(a) 5   **(b)** 4   **(c)** 3   **(d)** 2   **(e)** 1

8. In the figure shown, what is the value of x?

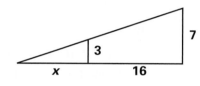

(a) 4   **(b)** 6   **(c)** 7   **(d)** 12   **(e)** 10

9. If $50 = 10^x$, then $x =$

(a) $\dfrac{10}{\sqrt{50}}$ **(b)** $\log 50$ **(c)** $\log 10$ **(d)** $\log 5$ **(e)** 10^5

10. Solve: $3\sqrt[3]{6x - 4} = -12$

(a) no real solutions   **(b)** $x = -4$ **(c)** $x = -10$

(d) $x = 3\frac{1}{3}$ **(e)** $x = 2$

The Geometry of Complex Numbers

In Unit 6, "Polynomial and Rational Functions," the imaginary number $i = \sqrt{-1}$ and complex numbers, $a + bi$, were introduced. In the process of investigating operations on complex numbers, you discovered and proved properties of complex number arithmetic that were remarkably similar to those for real numbers. You also learned a fundamental theorem of algebra: a polynomial function of degree n has exactly n complex number zeroes, including multiplicity.

In this lesson, you will investigate some of the geometry of complex numbers. You may recall from Unit 6 that every complex number can be defined by an ordered pair of real numbers:

$$a + bi \leftrightarrow (a, b)$$

This one-to-one correspondence permits representation of complex numbers as points in a coordinate plane, called the *complex number plane.*

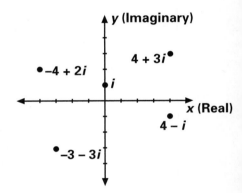

Think About This Situation

Geometric representations of mathematical ideas often provide insight and suggest direction for further possible developments.

a How does the representation of complex numbers as points in a coordinate plane suggest defining equality of complex numbers $a + bi$ and $c + di$?

b What possible connections do you see between complex numbers and vectors in a coordinate plane?

c How does the geometric representation of complex numbers suggest a way to define the absolute value $|a + bi|$ of a complex number?

d How would you geometrically interpret multiplying a nonzero complex number by a real number $r > 0$? By $r < 0$? By i?

e How could you represent a complex number $a + bi$ using polar coordinates?

INVESTIGATION 1 ▸ Complex Numbers and Vectors

The representation of complex numbers $a + bi$ as coordinates of points in a plane suggests a natural connection with vectors. In this investigation, you will explore that connection and the way it helps you think about complex numbers and operations on complex numbers from a geometric viewpoint.

1. The diagram at the right shows vector representations of four complex numbers. Note that the vectors are *position vectors* with initial point at the origin.

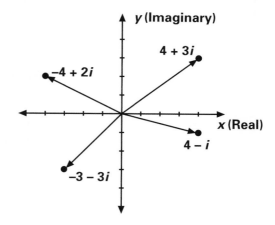

   a. Each member of your group should choose a different pair of complex numbers from the diagram. Then, using graph paper, verify that the sum of the two complex numbers can be represented as the sum of the corresponding vectors. Verify that the difference of the two complex numbers can be represented as the difference of the corresponding vectors.

   b. You may recall from Unit 6 that the *absolute value* of a complex number $|a + bi|$ is the length of the position vector (a, b). Which of the complex numbers in the diagram above has the greatest absolute value? In general, how would you calculate $|a + bi|$?

2. In working with absolute values of real numbers, one of the most important properties is the identity $|x + y| \le |x| + |y|$. Investigate whether this relationship also holds for complex numbers.

   a. Does the inequality $|z + w| \le |z| + |w|$ hold for the complex number sums your group calculated in Activity 1?

   b. Calculate the sums of the following pairs of complex numbers and check the inequality $|z + w| \le |z| + |w|$ in each case.

   i. $z = 1 + 2i$ and $w = 2.5 + 5i$

   ii. $z = 2 + i$ and $w = -3 - 1.5i$

   iii. $z = 4 - 5i$ and $w = -4 + 5i$

   c. For what pairs of complex numbers z and w is each of the following statements true?

   i. $|z + w| < |z| + |w|$?
   ii. $|z + w| = |z| + |w|$?

   d. Use a geometric argument with vectors to prove your answers in Part c.

3. Earlier in this unit, you learned that points in a plane can be described with polar coordinates, (r, θ), where r is the length of the position vector to the point and θ is the angle that the position vector makes with the positive x-axis.

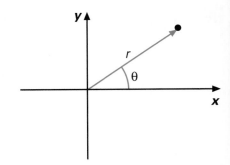

Find the rectangular coordinates and the complex number corresponding to each point with the given polar coordinates.

a. $\left(5, \frac{\pi}{6}\right)$ **b.** $\left(7, \frac{3\pi}{4}\right)$ **c.** $\left(3, -\frac{2\pi}{3}\right)$ **d.** $\left(r, \frac{2\pi}{3}\right)$ **e.** $(2, \theta)$ **f.** (r, θ)

4. You can use the relationships between the rectangular and polar coordinates of a complex number to express a complex number in *trigonometric form*. You will find this alternate representation useful in the remaining work in this unit.

   a. Suppose (r, θ) are the polar coordinates of a complex number $a + bi$. Write a and b in terms of r and θ. Rewrite $a + bi$ using these new expressions. This is the **trigonometric form** of $a + bi$.

   b. Express each complex number in standard form $a + bi$.

   i. $2 \cos \frac{\pi}{3} + 2i \sin \frac{\pi}{3}$ **ii.** $4\left(\cos \frac{7\pi}{6} + i \sin \frac{7\pi}{6}\right)$

   iii. $3\left(\cos \frac{\pi}{2} + i \sin \frac{\pi}{2}\right)$ **iv.** $5(\cos \pi + i \sin \pi)$

   c. Express each of the following complex numbers in trigonometric form.

   i. $-3i$ **ii.** $4 + 4i$

   iii. $-2 + 3i$ **iv.** $3 - 4i$

5. The trigonometric form of a complex number is especially useful in understanding the relationship between multiplication of complex numbers and geometric transformations in the plane such as rotations and size transformations.

   a. Consider first, the following two complex numbers:

   $$z = 5\left(\cos \frac{\pi}{4} + i \sin \frac{\pi}{4}\right) \text{ and } w = 3\left(\cos \frac{\pi}{2} + i \sin \frac{\pi}{2}\right)$$

   Sketch the vectors representing these two complex numbers. Then, calculate the product of the two complex numbers and sketch the vector representing that product. Write the product in trigonometric form.

   ■ Compare the absolute value of the product to the absolute values of the factors, and compare the angle of the product to the angles of the factors.

   ■ Let $v = 2\left(\cos \frac{3\pi}{4} + i \sin \frac{3\pi}{4}\right)$. Find vw. Write the product in trigonometric form. Do similar patterns among absolute values and angles occur?

b. Next consider, in general, two complex numbers:

$$z = r(\cos \alpha + i \sin \alpha) \text{ and } w = s(\cos \beta + i \sin \beta).$$

Supply reasons for each step in the following calculation for the product of these two numbers.

$$\begin{aligned} zw &= [r(\cos \alpha + i \sin \alpha)][s(\cos \beta + i \sin \beta)] \\ &= rs[(\cos \alpha + i \sin \alpha)(\cos \beta + i \sin \beta)] \\ &= rs[(\cos \alpha \cos \beta - \sin \alpha \sin \beta) + i(\cos \alpha \sin \beta + \cos \beta \sin \alpha)] \\ &= rs[\cos (\alpha + \beta) + i \sin (\alpha + \beta)] \end{aligned}$$

c. Explain what the result in Part b tells you about the multiplication of complex numbers in terms of a geometric transformation. What does it tell you in the case that each complex number has absolute value 1?

d. Now consider the connection between complex number arithmetic and rotations of a point about the origin. What complex number operation can be used to rotate a point $z = a + bi = r(\cos \theta + i \sin \theta)$ through an angle α about the origin? Explain your reasoning.

Checkpoint

In this investigation, you explored representations of complex numbers as points and as vectors in a coordinate plane.

ⓐ How are the sum and difference of two complex numbers represented in a diagram showing the vectors to which those numbers correspond?

ⓑ How do you calculate and interpret the absolute value of a complex number, $|a + bi|$?

ⓒ What is the trigonometric form of the complex number corresponding to the point with polar coordinates (r, θ)?

ⓓ How do you determine the trigonometric form of a complex number expressed in standard form $a + bi$?

ⓔ Write a summarizing statement about how to multiply complex numbers expressed in trigonometric form.

Be prepared to explain your ideas and methods to the entire class.

▶ On Your Own

Use connections between various representations of complex numbers in completing the following tasks.

a. Graph the following complex numbers in the complex number plane.

$$P: -5 + 2i \qquad Q: 2 - 4i \qquad R: 6 + i \qquad S: \left(5, -\frac{\pi}{6}\right)$$

b. Find the sum and difference of the complex numbers corresponding to points P and Q and make a sketch showing the corresponding vectors.

c. Using the correspondence between complex numbers and points:

- Calculate $|P|$, $|Q|$, and $|P + Q|$ and show both algebraically and geometrically why $|P + Q| < |P| + |Q|$.

- Find a complex number T for which $|T + R| = |T| + |R|$.

d. Write the complex number represented by point R in polar form.

e. Represent each complex number as a vector in the complex number plane.

$$z_1 = 2 + 5i \qquad z_3 = 4\left(\cos \frac{\pi}{2} + i \sin \frac{\pi}{2}\right)$$
$$z_2 = -6 + 3i \qquad z_4 = 3\left(\cos \frac{\pi}{6} + i \sin \frac{\pi}{6}\right)$$

- Multiply each complex number by i and represent the product as a vector. Describe geometrically the effect of multiplying a nonzero complex number by i.

- Investigate and summarize the geometric effect of multiplying a nonzero complex number by $-i$.

- Use complex number arithmetic to rotate the vector corresponding to z_3 counterclockwise $\frac{2\pi}{3}$ radians about the origin.

INVESTIGATION 2 DeMoivre's Theorem

Abraham DeMoivre

The connection between multiplication of complex numbers expressed in trigonometric form and rotations of the coordinate plane leads to a beautiful and useful principle known as DeMoivre's Theorem. That theorem, in turn, leads to solution of equations that go far beyond the simple quadratic equations that generated the need for complex numbers. By completing the activities in this investigation you will discover, prove, and learn how to apply the theorem attributed to Abraham DeMoivre (1667–1754), a French mathematician who is credited with many other important discoveries, especially in probability theory.

1. Consider the complex number $z = r(\cos \theta + i \sin \theta)$ and what you know about the connection between multiplication of complex numbers and rotations of the plane.

a. Write $z^2 = z \cdot z$ in trigonometric form. Interpret z^2 geometrically.

b. Using your result from Part a, write $z^3 = z^2 \cdot z$ in trigonometric form.

c. Extend the reasoning of Part b to write z^4, z^5, …, z^n in simplest possible trigonometric form. Use the pattern in these results to formulate a statement of the theorem attributed to DeMoivre: If $z = r(\cos \theta + i \sin \theta)$ and n is any positive integer, then $z^n =$ _____.

d. Use algebraic reasoning to demonstrate that

$r^{n-1}[\cos{(n-1)\theta} + i\sin{(n-1)\theta}] \cdot r(\cos\theta + i\sin\theta) = r^n(\cos{n\theta} + i\sin{n\theta})$.

This is the crucial step in proving DeMoivre's Theorem by mathematical induction. Complete a proof of DeMoivre's Theorem using the Principle of Mathematical Induction.

e. What does DeMoivre's Theorem imply about the trigonometric form of the complex number $z = \left[3\left(\cos\frac{\pi}{5} + i\sin\frac{\pi}{5}\right)\right]^{10}$? Explain why this number is actually equal to the real number 59,049.

2. DeMoivre's Theorem provides a simple way to calculate powers of any complex number when written in trigonometric form $z = r(\cos\theta + i\sin\theta)$. Reversing the reasoning suggested by that result provides a way of finding all complex number roots for polynomial equations in the form $z^n = a$, for any positive integer n. Those solutions, in turn, provide insight into some important geometric problems.

a. Consider first the cubic equation $z^3 = 1$. To find the roots, you need numbers which when raised to the third power give 1. Of course, one obvious solution is $z = 1$. But there are two others! Supply explanations for the steps in the mathematical argument below:

(1) The number $z = 1$ can be represented in trigonometric form as $(\cos 0 + i\sin 0)$, $(\cos 2\pi + i\sin 2\pi)$, $(\cos 4\pi + i\sin 4\pi)$, $(\cos 6\pi + i\sin 6\pi)$, and so on. So, you are seeking complex numbers $w = r(\cos\theta + i\sin\theta)$ so that $w^3 = r^3(\cos 3\theta + i\sin 3\theta)$ is equal to one of the trigonometric forms of 1.

(2) These conditions imply that $r = 1$.

(3) These conditions also imply that $\theta = 0$, $\theta = \frac{2\pi}{3}$, and $\theta = \frac{4\pi}{3}$ determine solutions. There are other values of θ that meet the given conditions, but they are all related to the basic three solutions by addition of multiples of the period 2π.

(4) The three values of θ obtained in step (3) determine the *complex roots* of the given equation. In standard form, the three solutions are

$$\cos 0 + i\sin 0 = 1$$
$$\cos\frac{2\pi}{3} + i\sin\frac{2\pi}{3} = -\frac{1}{2} + \frac{\sqrt{3}}{2}i$$
$$\cos\frac{4\pi}{3} + i\sin\frac{4\pi}{3} = -\frac{1}{2} - \frac{\sqrt{3}}{2}i$$

The complex roots of the equation $z^3 = 1$ are called the *cube roots of unity*.

b. The diagram at the right shows a circle of radius 1 in the complex number plane. On a copy of the diagram, draw point representations of the three cube roots of unity. Describe how the points are distributed on the circle.

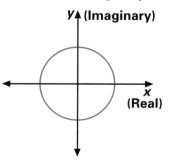

c. Now apply the same sort of reasoning to find the four *quartic roots of unity*; that is, find the four complex roots of the equation $z^4 = 1$. Plot these roots on a unit circle and describe their placement.

d. Describe a general procedure to find and represent the *n*th roots of unity.

e. Use reasoning similar to that in Part a to find the fifth roots of 243.

f. How would you find solutions for equations of the form $x^n = a$ for any positive power *n* and any nonzero real number *a*? In general, how would the point representations of these numbers in the complex number plane be related?

3. In Activity 2, you investigated methods used to find all roots of a given real number. In this activity, you will apply similar reasoning to find roots of imaginary numbers.

a. Consider the equation $z^3 = i$.

If $z = r(\cos \theta + i \sin \theta)$, then $z^3 = r^3(\cos 3\theta + i \sin 3\theta)$.

Since $i = \cos \left(\frac{\pi}{2} + 2\pi k\right) + i \sin \left(\frac{\pi}{2} + 2\pi k\right)$, where *k* is any integer, what can you conclude about *r*? About 3θ? About θ?

b. Use your conclusions in Part a to write the cube roots of *i* in trigonometric form.

c. Display the cube roots of *i* on a unit circle in the complex number plane. Compare their placement with the cube roots of 1.

d. Predict the locations of the cube roots of $-i$ on the unit circle. Check your conjecture.

4. You can extend your reasoning in Activity 3 to find the *n*th root of any nonzero complex number. Find the fourth roots of $-8 + 8\sqrt{3}i$ and display them on a circle in the complex number plane.

Checkpoint

In this investigation, you have explored the general problem of calculating the *n*th power of any complex number and the related problem of solving equations like $x^n = a$ for any positive integer *n* and any nonzero value of *a*.

ⓐ If $z = r(\cos \theta + i \sin \theta)$, then what is the trigonometric form of z^n?

ⓑ How does the connection between complex number multiplication and rotations of a coordinate plane explain the pattern in Part a?

ⓒ What are *n*th roots of unity, and how can their trigonometric and standard complex number forms be constructed using DeMoivre's Theorem? How can they be displayed in the complex number plane?

ⓓ Describe how to find the *n*th roots of $a + bi$.

Be prepared to explain your ideas and methods to the entire class.

Use DeMoivre's Theorem to help complete the following tasks involving powers and roots of complex numbers.

a. Plot each of the following complex numbers on a separate complex number plane, then calculate the indicated nth power and plot the result. Explain how the plot could be predicted from the geometry of rotations and size transformations.

   i. If $z = i$, find z^5.

   ii. If $z = 1 - i$, find z^3.

b. Find the angles corresponding to the seventh roots of unity and locate points corresponding to the roots on a unit circle in the complex number plane.

c. Find the angles corresponding to the seventh roots of -1 and locate points corresponding to the roots on a unit circle in the complex number plane. How is the polygon formed by connecting the seventh roots of unity related to the polygon found by connecting the seventh roots of -1?

MORE
Modeling • Organizing • Reflecting • Extending

Modeling

1. Sketch vector diagrams illustrating each of the following complex number operations.

   a. $(5 - 2i) + (2 + 4i)$

   b. $(5 - 2i) - (2 + 4i)$

   c. $3(-2 + 3i)$

   d. $\left[3\left(\cos \frac{\pi}{4} + i \sin \frac{\pi}{4}\right)\right]\left[2\left(\cos \frac{3\pi}{4} + i \sin \frac{3\pi}{4}\right)\right]$

2. Calculate each of the following absolute values and sketch diagrams showing the geometric length represented by each calculation.

   a. $|2 + 3i|$

   b. $|-5 - 4i|$

   c. $|(4 + 2i) - (2 - 3i)|$

   d. $|(4 + 2i) + (2 - 3i)|$

3. In your work in Unit 6, you saw that complex numbers have important applications in electronics and electrical engineering. The trigonometric form of complex numbers is particularly useful when dealing with applied problems involving both multiplication and division. Consider the *voltage divider formula* $V_1 = \dfrac{V_S Z_1}{Z_T}$ that is used to calculate the voltage V_1 across any element in an AC (alternating current) circuit. Here V_S is the applied voltage, Z_1 is the impedance of the element, and Z_T is the total impedance.

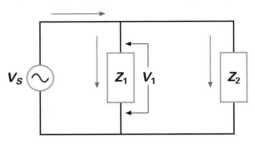

a. If $z = r(\cos \theta + i \sin \theta)$, what is the trigonometric form of $z^{-1} = \dfrac{1}{z}$? Explain your reasoning.

b. Use the fact that $\dfrac{w_1}{w_2} = w_1 \cdot \dfrac{1}{w_2}$ ($w_2 \neq 0$) to describe how to calculate $\dfrac{w_1}{w_2}$ when $w_1 = r_1(\cos \alpha + i \sin \alpha)$ and $w_2 = r_2(\cos \beta + i \sin \beta)$.

c. Use your result from Part b to calculate V_1, in the above diagram, if $V_S = 20(\cos 0 + i \sin 0)$ volts, $Z_1 = 75\left(\cos \dfrac{\pi}{4} + i \sin \dfrac{\pi}{4}\right)$ ohms, and $Z_T = 180\left(\cos \dfrac{\pi}{3} + i \sin \dfrac{\pi}{3}\right)$ ohms.

d. Express your voltage answer in Part c in standard form.

e. Determine the value of Z_1 in the voltage divider under conditions in which $V_1 = 12(\cos 10° + i \sin 10°)$ volts, $V_S = 28(\cos 0° + i \sin 0°)$ volts, and $Z_T = 200(\cos 75° + i \sin 75°)$ ohms.

4. One of the standard techniques of computer graphic design is to identify coordinates for key points of a figure and then produce the figure by connecting those points with line segments or curves. For example, the arrowhead shown at the right can be drawn by connecting four points with four segments.

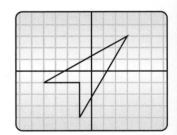

a. List the coordinates of key points that can be connected to form the figure. Write complex numbers corresponding to the key points.

b. Transform the arrowhead into a similar figure by multiplying each of the key points by $3\left(\cos \dfrac{3\pi}{4} + i \sin \dfrac{3\pi}{4}\right)$. Sketch the image figure. Describe geometrically the sequence of transformations that map the original figure to its image.

c. Find a complex number multiplier that transforms the arrowhead into an image with sides multiplied by 4 and leaves the tilt of the arrowhead and its image the same, that is, each side is parallel to its image side. Write the multiplier in both trigonometric and standard form. How would the multiplier be changed if sides were to be $\frac{1}{2}$ the original?

d. Suppose that some different figure is defined by a set of points with complex number representations $z = x + yi$. What complex number operation will transform each such point so that the resulting figure has sides that are m times as long as and parallel to their pre-images? Use the fact that the distance between any two points $z = x + yi$ and $w = s + ti$ can be calculated from $|z - w|$ to prove that your rule works.

5. Find the following powers of complex numbers. In each case, determine the absolute value and the quadrant in which the corresponding point lies.

   a. $[3(\cos \pi + i \sin \pi)]^5$ b. $\left[5\left(\cos \frac{\pi}{2} + i \sin \frac{\pi}{2}\right)\right]^7$

   c. $\left[2\left(\cos \frac{\pi}{3} + i \sin \frac{\pi}{3}\right)\right]^4$ d. $(-1 + i)^8$

6. Find all specified roots of the given complex numbers and express roots in standard form. In each case, sketch the roots on a circle in the complex number plane.

   a. Sixth roots of 1   b. Sixth roots of -1

   c. Cube roots of $-27i$ d. Fourth roots of -81

   e. Fifth roots of $32i$

Organizing

1. Recall from your work in Unit 6 that, for any complex number $z = a + bi$, the number $a - bi$ is called the *conjugate* of z. The notation $\bar{z}$ (read "z bar") is used to denote the conjugate of z.

   a. What geometric transformation maps each complex number onto its conjugate?

   b. Show that for any complex number $z, |z| = |\bar{z}|$.

   c. Show that for any complex number z, the product $z\bar{z} = |z|^2$.

   d. Show that for any complex numbers z and w, $\overline{z + w} = \bar{z} + \bar{w}$. What property of real numbers is similar to this property of complex conjugates?

   e. If $r + si$ is one solution of a quadratic equation $ax^2 + bx + c = 0$ with real coefficients, what is the other solution? Explain your reasoning.

2. Solve the equation $z \cdot \bar{z} + 2(z - \bar{z}) = 10 + 6i$ over the set of complex numbers.

3. Recall that if $z = r(\cos \alpha + i \sin \alpha)$ and $w = s(\cos \beta + i \sin \beta)$,

$$z \cdot w = rs[\cos (\alpha + \beta) + i \sin (\alpha + \beta)].$$

   a. If a figure is defined by a set of n points with complex number representations $w_k = s_k(\cos \beta_k + i \sin \beta_k)$ for $k = 1, 2, \ldots, n$, what is the effect on the figure if each w_k is multiplied by z? Demonstrate your idea with calculations and sketches in the case of $z = 2.5\left(\cos \frac{\pi}{4} + i \sin \frac{\pi}{4}\right)$ by locating the image of the arrowhead in Modeling Task 4.

b. What sequence of transformations could be used to transform the arrowhead as in Part a? What is the effect of the composite transformation on lengths, on shape, on area, and on tilt.

c. Use the fact that the distance between $w_1 = x + yi$ and $w_2 = s + ti$ is $|w_1 - w_2|$ to prove that multiplication by $z = r(\cos \alpha + i \sin \alpha)$ multiplies distances by r. (The transformation performed by multiplying by z is called a *spiral similarity*. It is the composite of a size transformation with magnitude r and a rotation.)

4. Consider the functions with rules of the form $f(z) = Az + B$, where z is a complex number variable and A and B are specific complex numbers. What sort of geometric transformation will be determined in each of the following cases?

a. $A = 1$

b. $B = 0$ and $|A| = 1$

c. $B = 0$ and A is a positive real number different from 1

5. Prove that if $z^n = r(\cos \theta + i \sin \theta)$ where n is a positive integer, then the nth roots are given by $z = \sqrt[n]{r}\left[\cos\left(\frac{\theta}{n} + \frac{2\pi k}{n}\right) + i \sin\left(\frac{\theta}{n} + \frac{2\pi k}{n}\right)\right]$, where k is any integer.

Reflecting

1. Addition and scalar multiplication of vectors are very similar to addition of complex numbers and multiplication of a complex number by a real number. What does complex number arithmetic add to the geometry that is not part of your previous vector work?

2. Recall that for any nonzero real number r, $r^0 = 1$. Reasoning analogously, define $z^0 = 1$ for any nonzero complex number. Verify that DeMoivre's Theorem also holds for $n = 0$.

3. In the 18th century, the Swiss mathematician Leonhard Euler discovered a surprising connection between the expression e^x that you studied in Lesson 1 and the expressions $\sin x$ and $\cos x$. The connection involves complex numbers. **Euler's formula** states that $e^{i\theta} = \cos \theta + i \sin \theta$, provided that θ is measured in radians.

a. Evaluate $e^{i\pi}$.

b. The statement you wrote, known as *Euler's identity*, is amazing in that it relates five of the most fundamental constants in mathematics. What are they?

c. Euler is depicted on the German stamp issued in his honor shown at the left. The stamp shows another formula for which Euler is famous, $e - k + f = 2$. You may have discovered that formula in a previous course. What relationship is represented by the formula?

4. What trigonometric identities turn out to be critical in relating complex number multiplication to geometric rotations and DeMoivre's Theorem?

5. If the nth roots of a complex number $z = r(\cos \theta + i \sin \theta)$ are graphed on a circle in the complex number plane, what can you say about the radius of the circle? About the measure of the angle formed by the vector representations of two consecutive roots?

Extending

1. Supply justifications for each step in the following algebraic proof of DeMoivre's Theorem in the case $n = 2$:

$$[r(\cos \theta + i \sin \theta)]^2 = r^2[(\cos \theta + i \sin \theta)(\cos \theta + i \sin \theta)]$$
$$= r^2[(\cos^2 \theta - \sin^2 \theta) + i(2 \sin \theta \cos \theta)]$$
$$= r^2(\cos 2\theta + i \sin 2\theta)$$

2. Use DeMoivre's Theorem to derive identities for $\cos 3\theta$ and for $\sin 3\theta$ in terms of $\cos \theta$ and $\sin \theta$.

3. The nth roots of unity all have absolute value equal to 1. That means that they correspond to points on a unit circle in the complex number plane.

  a. Construct a table showing the angles (in radians) that determine roots of unity for $n = 3, 4, 5, 6, 7,$ and 8. Inspect the results and find a formula for the angles that give nth roots of unity for any n.

  b. What sort of polygon is formed when the nth roots of unity are connected in order of increasing angle (starting at $\alpha = 0$)? Explain how you know that your answer is correct by relating the pattern to the definition of radian measure for angles.

4. Consider all functions $f(z) = Az + B$, where z is a complex number variable and A and B are specific complex number constants.

  a. Show that the composite of any two such functions is another function of the same type.

  b. Show that whenever $|A| = 1$, the resulting function preserves distances between pairs of complex numbers, that is, show that $|f(z_1) - f(z_2)| = |z_1 - z_2|$.

  c. Show that if $A \neq 1$, the function has exactly one fixed point, that is, there is exactly one value of z for which $f(z) = z$. Explain how that allows you to conclude that the function corresponds to a spiral similarity of the plane about the fixed point. (See Organizing Task 3.)

  d. Show that if $A = 1$ and $B \neq 0$, the function has no fixed points. Explain how that allows you to conclude that the transformation is a translation of the plane.

  e. Explore similar cases for functions of the form $g(z) = A\bar{z} + B$, and see what you can conclude about the transformation of the plane described by such functions for various combinations of A and B.

1. $\left(3 \div \frac{1}{9}\right) + \left(9 \div \frac{1}{3}\right) =$

   (a) $\frac{10}{3}$ (b) 30   (c) $\frac{2}{27}$ (d) 54   (e) 36

2. Simplify: $3x^2(x - 5x^3) - 4x^3(x^2 - 7)$

   (a) $12x^2$ (b) $31x^6 - 19x^{10}$ (c) $31x^3 - 19x^5$

   (d) $-25x^3 - 19x^5$ (e) $31x^3 - 19x^6$

3. An equation of the line containing $(5, 1)$ and $(0, -6)$ is

   (a) $y = \frac{7}{5}x + 6$ (b) $y = \frac{7}{5}x - 6$ (c) $y = -\frac{7}{5}x - 6$

   (d) $y = \frac{5}{7}x - 6$ (e) $y = \frac{5}{7}x + 6$

4. If $f(x) = x^2 + 2$ and $g(x) = x + 4$, then $f(g(x)) =$

   (a) $x^2 + x + 6$ (b) $2x^2 + 1$ (c) $x^2 + 8x + 18$

   (d) $6x$ (e) $x^2 + 6$

5. Solve over the complex numbers: $x^2 + 9 = 0$

   (a) $x = 3i$ or $x = -3i$ (b) $x = 3$ (c) $x = 3$ or $x = -3$

   (d) $x = -3$ (e) $x = 3 + i$ or $x = 3 - i$

6. Solve: $\frac{x + 6}{4} - \frac{x - 3}{7} < 9$

   (a) $x < -15$ (b) $x < 102$ (c) $x < 74$ (d) $x > 15$ (e) $x < 66$

7. Which of the following could be a portion of the graph of $y = \dfrac{(x+1)(x-6)}{x-3}$?

(a) 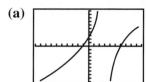　　(b) 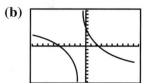　　(c)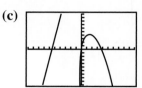

(d) 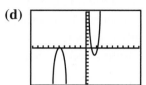　　(e)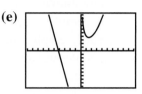

8. If a square is inscribed in a circle with a radius of 1 unit, what is the area of the square?

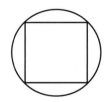

(a) 4　　　(b) 2　　　(c) $2\sqrt{2}$　　(d) $\dfrac{1}{\sqrt{2}}$　　(e) 1

9. $\log_3 81 =$

(a) 27　　(b) $\dfrac{3}{81}$　　(c) 4　　(d) $\dfrac{1}{4}$　　(e) $3\sqrt[3]{3}$

10. Which of the following could be the equation for the graph below?

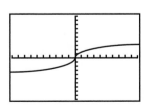

(a) $y = x^{\frac{1}{3}}$　　(b) $y = x^3$　　(c) $y = x^{\frac{1}{2}}$　　(d) $y = x^2$　　(e) $y = x$

Lesson 5

Looking Back

In this unit, you have revisited and deepened your understanding of exponential, logarithmic, and trigonometric functions and applied the latter to further your study of complex numbers. Whereas your earlier experiences with these functions primarily emphasized verbal, graphical, and numerical representations, the focus of this unit was on transforming and reasoning with the corresponding symbolic representations. Through combinations of symbolic reasoning and manipulation strategies, you were able to derive equivalent and, often, more easily interpreted or used forms of symbolic expressions. In the case of the trigonometric functions, this work led to the derivation of identities useful in advanced mathematics and in physics.

As you complete the review and summarizing tasks that follow, think about how reasoning with and manipulating symbolic expressions enhances your understanding of mathematical models and their uses. Consider also the precision and economy of thought (and work) that is often gained through symbolic reasoning. Reflect on how you could check your solutions by reasoning with tables and graphs.

1. Suppose that in a laboratory an infectious bacteria colony was isolated and an experimental antibiotic was applied. The number of bacteria as a function of time t in hours was modeled by $h(t) = 1,000(0.0625)^{\frac{t}{8}}$.

   a. How many bacteria were present at the time the antibiotic was applied?

   b. Is the function $h(t)$ an increasing or decreasing function? Describe the rate of growth or decay.

   c. Determine the hourly rate of growth or decay.

   d. Express $h(t)$ in an equivalent form with base one-fourth. Interpret this representation.

   e. Express $h(t)$ in an equivalent form with base e. What is an advantage of this representation?

   f. What is the half-life of the bacteria under this treatment?

   g. Is $h(t) = 10^{3 - 0.1505t}$ equivalent to $h(t) = 1,000(0.0625)^{\frac{t}{8}}$? Justify your answer using symbolic reasoning.

2. Recall from your work in Unit 3, "Logarithmic Functions and Data Models," that the decibel scale for sound intensity is a logarithmic scale. When sound intensity I is measured in watts per square centimeter, the formula for calculating decibels is $D = 10 \log (10^{16} \cdot I)$.

   a. The intensity of a whisper is about 10^{-13} watts/cm^2. How many decibels is this?

   b. Sound produced by traffic has intensity about 80 decibels. Find the intensity of the sound vibrations in watts per square centimeter.

   c. Two sounds differ by about 3 decibels. How do the intensities of the two sounds compare?

   d. The intensity of one sound is 5 times that of another. How do the decibel readings for the two sounds compare?

3. Solve each equation.

   a. $200(0.886^x) = 25$

   b. $2.3(1.76^{x-3}) - 2 = 7$

   c. $95e^{-0.112x} = 80$

   d. $3 \log (2x + 6) = 6$

   e. $\ln (x - 3) + \ln (x + 3) = 1$

4. As you have seen previously, the number of minutes between sunrise and sunset in any one location can be modeled by a trigonometric function. A modeling rule for this number for Washington, D.C. is $S(d) = 180 \sin (0.0172d - 1.376) + 720$, where d is the day of the year with January 1 as $d = 1$.

   a. On what day is the amount of time between sunrise and sunset closest to 900 minutes?

   b. On which days during a year will there be more than 12 hours between sunrise and sunset in Washington, D.C.?

   c. Will there ever be a day where there are less than 8 hours between sunrise and sunset in Washington, D.C.?

5. Suppose that you are observing the motion of a clock pendulum. The distance the pendulum is to the left or right of vertical can be modeled by a cosine function.

For one clock, the modeling rule is $d(t) = 5 \cos \pi t$, where t is the number of seconds since the pendulum was released at the right endpoint of its swing.

a. How long does it take for one complete swing of the pendulum?

b. At what times will the pendulum be exactly vertical?

c. During what time periods will the pendulum be to the left of vertical?

6. The amount of power that a city uses varies with the time of day. Suppose that the power requirements in megawatts (MW) for Santa Rio are modeled by the function $P(t) = 40 - 20 \cos \left(\frac{\pi}{12} t - \frac{\pi}{4} \right)$, where t is the number of hours after midnight on Sunday night.

a. Explain why this is a reasonable model. Over what domain do you think this modeling equation will be valid?

b. At what times will the city need 40 MW of power?

c. Under present conditions on power demands, will the city ever need 80 MW of power? Explain.

d. During what time periods will the city need less than 30 MW of power?

7. Prove each trigonometric identity.

a. $(1 - \tan \theta)^2 = \sec^2 \theta - 2 \tan \theta$

b. $\frac{\cot \theta}{\cos \theta} + \frac{\sec \theta}{\cot \theta} = \sec^2 \theta \csc \theta$

c. $\frac{\sec^2 \theta}{\sec^2 \theta - 1} = \csc^2 \theta$

d. $(\sin \theta + \cos \theta)(\tan \theta + \cot \theta) = \sec \theta + \csc \theta$

e. $\cos (\alpha - \beta) \cos (\alpha + \beta) = \cos^2 \alpha - \sin^2 \beta$

f. $2 \sin \left(\frac{\pi}{4} + \alpha \right) \sin \left(\alpha - \frac{\pi}{4} \right) = \sin^2 \alpha - \cos^2 \alpha$

g. $\tan^2 \theta = \sec^2 \theta - \sin^2 \theta - \cos^2 \theta$

h. $\tan \alpha = \frac{1 - \cos 2\alpha}{\sin 2\alpha}$

8. Solve each equation on the interval $-2\pi \le x \le 2\pi$.

   a. $4 \sin^2 x = 1$

   b. $2 \tan^2 x - 3 \sec x + 3 = 0$

   c. $\sin^2 x - \cos^2 x - \cos x - 1 = 0$

   d. $6 \tan^2 x + 5 \tan x + 1 = 0$

9. Shown below are computer algebra system (CAS) solutions to an exponential equation and a trigonometric equation.

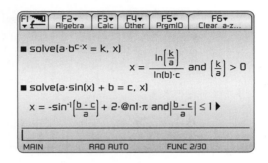

   a. Explain why the CAS solution of the general exponential function $a(b^{cx}) = k$ makes sense.

   b. The marker ▶ on the right of the display of a solution of $a \sin(x) + b = c$ indicates that more follows. What other solutions can be expected?

10. A square has its vertices at $(2, 2)$, $(-2, 2)$, $(-2, -2)$, and $(2, -2)$.

    a. On a coordinate system, sketch the square and write the complex numbers that correspond to the vertices:

       ■ in standard form

       ■ in polar form

       ■ in trigonometric form

    b. Express $1 - i$ in polar form and in trigonometric form. Then find the image of the square in Part a under the transformation accomplished by multiplying each vertex by $1 - i$. Describe the effects of this transformation on the square.

    c. $4 + 0i$ and $0 + 4i$ are vertices of the image figure in Part b. Find the square roots of each point.

This unit focused on methods for reasoning with and manipulating symbolic representations involving exponential, logarithmic, and trigonometric functions in both applied and mathematical contexts.

a What are the fundamental properties of exponents and of logarithms that are used to rewrite exponential and logarithmic expressions in equivalent forms?

b Compare the natural logarithm function and the common logarithm function in terms of their definitions and in terms of their graphs.

c Explain why an exponential function $f(x) = a(b^x)$ can always be expressed as $f(x) = a(e^{kx})$. What are advantages of this alternative representation?

d Describe a symbolic reasoning strategy for solving $a(b^{cx + d}) = k$ for x. Note any restrictions that must be made so that a solution exists.

e What does it mean to prove an identity? What are the fundamental relationships that are most useful in proving identities?

f How are the sine and cosine of the sum of two angles related to the sines and cosines of the individual angles? What about the sine and cosine of the difference of two angles?

g Describe a symbolic reasoning strategy for solving $a \sin (cx + d) = k$ for x. Note any restrictions that must be made so that a solution exists.

h How is the trigonometric form of a complex number derived from the standard form of the number, $a + bi$?

i Explain how geometric transformations are connected to complex numbers and their operations.

j What is DeMoivre's Theorem, and how is it useful in finding roots of a complex number?

Be prepared to explain your responses to the entire class.

▶On Your Own

Write, in outline form, a summary of the important mathematical concepts and methods developed in this unit. Organize your summary so that it can be used as a quick reference in future units and courses.

Space Geometry

Representing Three-Dimensional Shapes

Hikers regularly use *contour maps* of trails to get an overall picture of the terrain, showing where the mountains are and where the flat regions are. The display below shows a contour map for the Buena Vista Trail of the White Mountains in Apache-Sitgreaves National Forest, Arizona. The trailhead is at the star. The trail is the light blue dashed line. The light solid lines are *contour lines*. If a hiker walks along a contour line, she stays at the same elevation. On this map, adjacent contour lines differ in elevation by 40 ft.

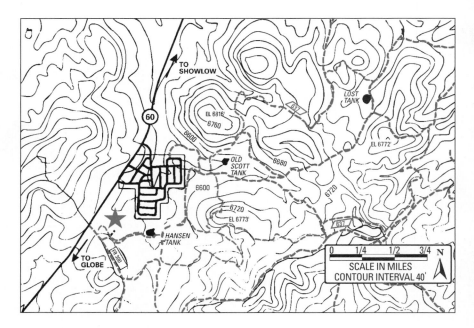

Contour maps are one way to represent three-dimensional shape in two dimensions. Other methods you have previously studied include perspective drawings, face-views, and isometric drawings. In this unit, you will learn how to represent and analyze three-dimensional shapes using contour diagrams, using cross sections, and using coordinates and equations.

INVESTIGATION 1 ▶ Contour Diagrams

In previous work, you constructed models of space-shapes and represented them with drawings such as those shown in Activity 1.

1. Examine each of the following diagrams of a space-shape. Note how the diagrams give the impression of depth, as well as of width and height.

 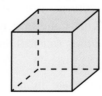

a. Identify each space-shape as completely as possible.

b. In the cases of right prisms:
   - When are congruent segments drawn the same length. When are they not? Why?
   - When are rectangular faces drawn as rectangles? When are they not?

c. Without tracing, make careful sketches of the diagrams above. Compare your procedure to those of other members of your group.

d. Make a careful sketch of a right rectangular prism that is not a cube.

2. Suppose that each space-shape of Activity 1 is resting on its base on a flat surface and that the height of each space-shape is 8 inches.

  a. Imagine horizontal planes slicing each space-shape at 2-in. vertical intervals. For each space-shape, describe the *cross sections* formed by the intersections of the planes with the surface of the shape.

  b. The diagram on the right shows how the series of cross sections in Part a for the rectangular pyramid can be combined to form a *contour diagram*. For each of the other space-shapes draw a similar contour diagram. How are the various contour diagrams alike and how are they different?

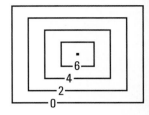

3. Now carefully sketch a circular cone and a triangular pyramid with each shape resting on its base.

  a. Assuming that the height of each shape is 8 in. and that contour lines are based on horizontal cross sections at 2-in. intervals, draw the corresponding contour diagrams.

  b. How would the contour diagrams for the cone and pyramid change if each shape were resting on the vertex instead of the base?

Just as rectangles are often replaced by nonrectangular parallelograms to give the impression of depth in two-dimensional drawings, circles are often replaced by ovals or ellipses.

4. Consider the sphere shown below. You know that each cross section of a sphere is a circle.

  a. Which numbered circular cross sections are drawn as circles? Why?

  b. Which numbered circular cross sections are congruent? Under what condition will two horizontal or vertical cross sections be congruent?

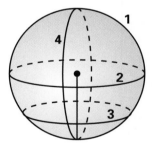

  c. Suppose the diameter of the sphere is 10 cm and that horizontal cross sections are taken every centimeter. Sketch the corresponding contour diagram.

  d. How would a contour diagram of the bottom hemisphere of the sphere differ from the contour diagram of the sphere? From the contour diagram of the top hemisphere?

  e. How does the contour diagram for a sphere differ from that of a cone standing on its base?

5. Assuming that the height of each space-shape below is 8 cm, sketch the corresponding contour diagrams for cross sections every 2 vertical centimeters.

a.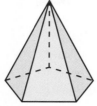

b.

c.

d.

Checkpoint

In this investigation, you reviewed how to represent three-dimensional shapes with two-dimensional drawings and explored the use and construction of contour diagrams.

a Describe how to make a contour diagram of a space-shape.

b What does it mean if contour lines of a surface are close together?

c Describe the advantages and disadvantages of representing three-dimensional shapes with sketches; with contour diagrams.

Be prepared to share your methods and ideas with the entire class.

▶On Your Own

Imagine or sketch a right circular cone resting on its base.

a. Suppose two horizontal planes slice the cone in two places other than the vertex. What is the shape of each cross section?

b. Sketch the part of the cone between the two planes in Part a.

c. If the height of the cone is 15 in. and cross sections are taken every 3 in., draw the contour diagram for the cone.

INVESTIGATION 2 ▶ Using Data to Determine Shape

NASA (the National Aeronautics and Space Administration) has sent men to the Moon and spacecrafts to Mars and Venus. In preparing to land on the Moon, a critical concern was identifying a safe location for landing the spacecraft. In the case of our moon, the surface was visible, so relatively flat sites could be located.

However, in seeking a flat landing site on Venus, the problem was more difficult because the surface was obscured by clouds. Orbiting NASA spacecraft used radar to determine the elevations of many points on the surface of Venus. These elevations were then used to make contour maps of portions of the surface. The maps were used to identify possible landing sites. During the 1978 Pioneer mission to Venus and the 1990 Magellan mission, nearly the entire surface of Venus was mapped using *radar altimetry*.

Exploration 1

You can simulate the radar altimeter mapping of Venus by using a partially-filled shoe box to model a region of the surface of the planet and a bamboo skewer as a radar probe. You will also need centimeter graph paper, a centimeter ruler, and a marking pen.

Calibrate your radar probe by making a mark 1 cm from the pointed end. Continue marking centimeter intervals on the entire skewer. On graph paper, prepare a grid corresponding to the one on the box top. Assume that the bottom and left edges of the shoe box (see photo at the right) correspond to the positive *x*- and *y*-axes, respectively.

1. You are now ready to collect data.
   Keep the shoe box closed. Do not look at the terrain inside.

   i. Insert the radar probe perpendicularly into a hole on the shoe box top. Continue the gentle insertion until it stops. Note the coordinates of the hole and the closest centimeter mark above the box top.

   ii. Extract the probe and identify the number of the centimeter mark determined in Part i. Record this number on the graph paper at the coordinates (x, y) that correspond to those on the box top hole.

   Repeat Parts i and ii for each box top hole.

2. Now, using your data, make a contour map of the terrain. Follow the procedure illustrated below. Do not open the shoe box and look at the terrain inside.

   i. Find a peak or valley. Draw a loop around those numbers in your data. It is best to start with either the highest peak or lowest valley.

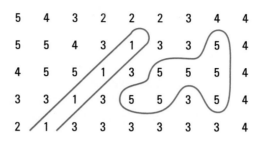

In the diagram above, a contour is drawn around a collection of 5s. Also an open contour is drawn around the 1s. It is open because there may be more 1s beyond the edge.

   ii. Work outward from the chosen peak or valley toward the edges of the map. For example, if 1 and 3 are next to each other, somewhere between them there is probably a height of 2. Why? So, starting with the 1-peak, you can draw the 2-contour. Note that the 1-peak is interior to the 2-contour.

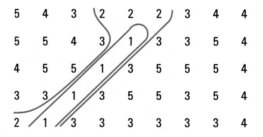

   iii. Continue with the remaining contours as illustrated below.

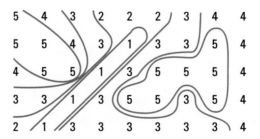

3. Now examine your contour map.

   a. How is a region of small numbers related to a region of larger numbers? Explain why this makes sense.

   b. Should the loops drawn for two numbers intersect? Explain.

   c. Based on your data, describe the terrain in the box.

   d. Now open the box and compare your description of the terrain to the actual surface.

Exploration 2

The following altitude data were collected by the Pioneer mission in 1978. The location is a region near the Venusian equator. For the purpose of this exploration, altitude data are given every 90 feet and then rounded to the nearest multiple of 10 ft.

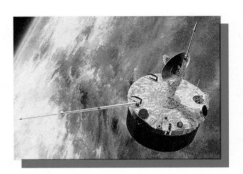

1. Examine these data and then on a copy of the data draw a contour map with contours at 10-ft altitude intervals starting at 690.

720	700	690	700	720	780	820	830	830	830	830	840	840	840	850	850	850
700	710	710	720	750	800	820	830	830	830	830	830	840	840	840	850	850
690	720	740	760	790	820	830	830	830	830	820	830	830	840	840	840	840
700	730	760	800	820	830	830	820	820	810	810	820	830	840	840	840	840
720	740	780	820	840	840	830	810	800	800	800	810	830	840	840	840	840
740	750	790	820	840	840	820	800	790	780	780	800	820	840	850	850	860
780	790	800	820	830	820	800	780	770	770	770	790	820	850	860	870	890
800	800	820	810	800	790	780	770	770	770	770	790	820	860	890	910	930
800	810	810	790	780	770	760	760	770	770	770	790	820	860	910	940	970
770	780	780	770	760	760	760	760	770	770	780	790	820	860	920	960	980
750	750	740	750	750	760	760	760	770	780	790	790	810	860	920	970	990
740	740	740	740	740	750	760	770	780	790	790	800	810	850	910	960	980
740	740	740	740	750	750	760	770	780	790	800	800	800	840	890	940	980
740	740	750	750	750	760	770	780	790	790	800	800	810	830	880	930	970
740	740	740	740	740	760	770	780	780	790	800	800	800	820	860	910	960
750	740	740	730	740	750	770	780	780	790	800	800	810	820	850	900	960
750	750	730	730	740	760	780	790	790	800	810	810	810	820	840	890	950

Source: http://nssdc.gsfc.nasa.gov

2. Using your contour map, locate and justify a possible landing site for a space probe.

3. A portion of the altitude data in 90-foot intervals from the lower left-hand corner of the surveyed region on Venus is shown on the coordinate system below. How could you describe each of the following lettered locations on the planet using three numbers (coordinates)?

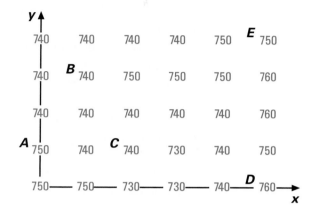

a. Point A

b. Point B

c. Point C

d. Point D

e. Point E

Checkpoint

In this investigation, you examined how to draw contour maps from altitude data and considered how to use three coordinates to identify locations in space.

ⓐ Describe how contour lines can be constructed for a three-dimensional surface.

ⓑ Describe how three numbers can be used to specify the location of a point in *three-space* (three-dimensional space).

ⓒ Using three coordinates, describe the location of a possible landing site for a spacecraft mission to Venus.

Be prepared to share your group's ideas with the entire class.

The three-number descriptions of a point in three-dimensional space used in this investigation are rectangular coordinates of the point. Rectangular coordinates of a point are determined by measuring the directed perpendicular distance from the point to each of three mutually perpendicular planes called *coordinate planes*. The *origin* is the intersection of the three coordinate planes. The intersections of the pairs of planes are the *coordinate axes*. The axes are usually called the *x*-, *y*-, and *z*-axis. The labels *x*, *y*, and *z* indicate the positive direction on each axis. You will explore more formally how to use three-dimensional coordinates in the MORE set and in the next lesson.

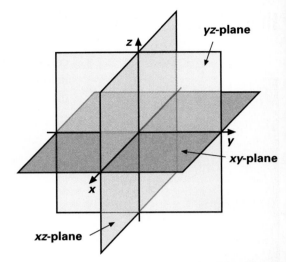

▶On Your Own

The height (in feet) of a surface, represented by the *z*-values below, was measured for the 49 lattice points in a square region: $-3 \leq x \leq 3$, $-3 \leq y \leq 3$. Use these data to construct a contour map for the height *z* above the *xy*-plane. Draw contour lines at 0.5-ft intervals. Look for patterns in the data that might reduce the effort and time in plotting the data. Use your contour map to describe the shape of the surface mapped.

x\y	−3	−2	−1	0	1	2	3
−3	7	9.5	11	11.5	11	9.5	7
−2	9.5	12	13.5	14	13.5	12	9.5
−1	11	13.5	15	15.5	15	13.5	11
0	11.5	14	15.5	16	15.5	14	11.5
1	11	13.5	15	15.5	15	13.5	11
2	9.5	12	13.5	14	13.5	12	9.5
3	7	9.5	11	11.5	11	9.5	7

INVESTIGATION 3 Visualizing and Reasoning with Cross Sections

Contour maps are developed by sampling heights at specific intervals. The procedure only approximates the true characteristics of the terrain. However, for many people who use contour maps, such as hikers, forest rangers, deep sea divers, and ship navigators, the approximate nature of the information is adequate. They want to know generally what the land or sea bottom near them looks like. For a hiker taking the Country Club Trail shown below, it is usually enough to know the summit of Pat Mullen Mountain is about 300 ft above the main trail and whether there are cliffs or valleys along the trail.

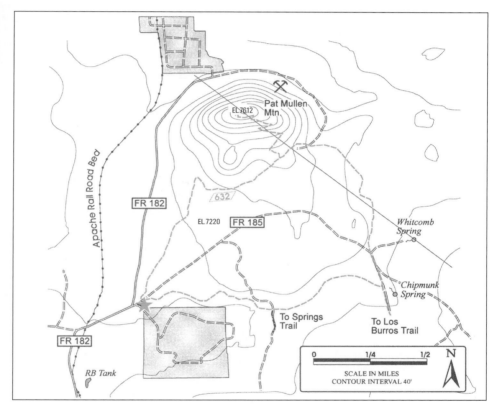

A contour map can be used to create a *relief map* of a region that also shows the variation in the altitude of the land surface. A relief map is made by imagining a vertical cross section of the terrain represented by a contour map.

1. Consider the vertical cross section of the region around Pat Mullen Mountain indicated by the straight line in the contour map above. It runs from Whitcomb Spring through the top of the mountain. A line like this is called a *relief line*.

   a. If the contour line at the summit of Pat Mullen Mountain is 7,600 ft and the contour lines are at 40-ft elevation intervals, what is the elevation at Whitcomb Spring?

b. Along the relief line, what can you say about the land surface if the contour lines are close together? If they are far apart?

c. Make a grid like that below. Measure the distance from one contour line to the next and plot the elevation versus the horizontal distance of the contour line from the summit of Pat Mullen Mountain. The first point to the left of Whitcomb Spring is plotted. Connect the points to obtain the relief map.

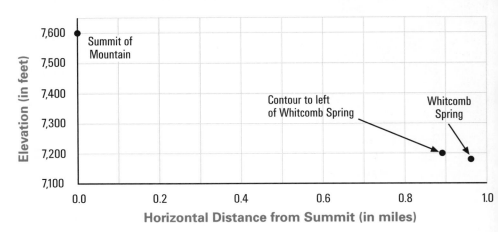

d. Describe how the spacing between the contour lines is represented on the relief map.

e. Make a relief map for the line indicated on the Ice Cave Trail map below.

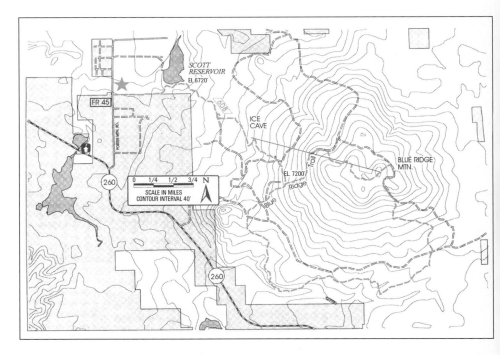

Just as a contour map can help you visualize portions of the surface of Earth or other planets, cross sections of mathematical shapes can reveal important features of those shapes.

2. Consider a cylinder with its base in the *xy*-plane as shown below.

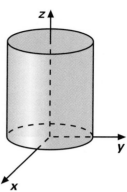

  a. Describe the cross sections of the cylinder if the intersecting planes are parallel to the *xy*-plane.

  b. Could any other space-shape have cross sections parallel to its bases exactly like those you described in Part a? If so, describe the shape.

  c. Describe cross sections of the cylinder if the intersecting planes are parallel to the *yz*-plane.

  d. How could you display or record both the horizontal and vertical cross sections so that they could be used to identify and sketch the cylinder? Compare your method to that of other groups.

3. The *circular paraboloid* below is the three-dimensional analog of a parabola. It was formed by rotating a parabola with vertex at the origin about the *z*-axis.

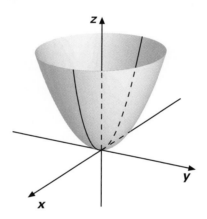

  a. Sketch five cross sections of this shape determined by equally-spaced planes parallel to the *xy*-plane.

  b. Repeat Part a for planes parallel to the *yz*-plane.

  c. How could you label and display the cross sections so that they could be used to identify and sketch the shape?

The display you prepared in Activity 3 is similar to a contour diagram in that a series of horizontal cross sections are determined by a series of parallel planes at uniform intervals. It is standard practice to record the series of horizontal cross sections with the one corresponding to the highest point drawn first and subsequent cross sections placed to the right. For the series of vertical cross sections, the first-drawn cross section is the one made by the vertical plane nearest the viewer; subsequent cross sections will be those obtained by intersections with vertical planes at uniform intervals further from the viewer.

4. Consider the following ordered horizontal and vertical cross sections of three-dimensional shapes. Use the cross sections to help you describe the actual shape.

  a. Horizontal cross sections:

Vertical cross sections:

  b. Horizontal cross sections:

Vertical cross sections:

  c. Vertical cross sections for another shape with the horizontal cross sections in Part a:

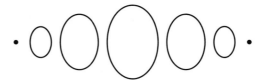

  d. Horizontal cross sections:

Vertical cross sections:

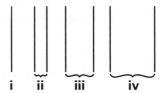

In this investigation, you explored how relief maps are created from contour maps and how analysis of both vertical and horizontal cross sections of a surface can help you better understand the surface.

ⓐ Explain how to make a relief map.

ⓑ How is a relief map similar to, and different from, a vertical cross section of a surface?

ⓒ Explain how cross sections of a three-dimensional shape can be used to figure out what the shape looks like and how to sketch it.

Be prepared to share your ideas with the entire class.

On Your Own

Horizontal cross sections of a space-shape are given at the right. Cross sections are made at 2-cm intervals above and below the middle cross section.

a. Sketch a shape that has the given horizontal cross sections.

b. Make sketches of several vertical cross sections of the shape.

INVESTIGATION 4 Conic Sections

Cross sections of space-shapes are plane shapes. As you observed in previous investigations, different plane shapes may result when planes and a shape in space intersect. An important class of plane shapes are those that can be obtained from the intersection of a *double cone* and a plane as shown below. Such intersections are called **conic sections**, or simply **conics**.

Circle	**Ellipse**	**Parabola**	**Hyperbola**
Plane intersecting one cone parallel to a base	Plane intersecting one cone not parallel to a base	Plane intersecting one cone parallel to edge of cone	Plane intersecting both cones perpendicular to a base

The various intersections of a plane with the surface of a double cone are familiar curves. In Unit 2, "Modeling Motion," you used parametric equations of circles to model the motion of rotating CDs, you used parametric equations of ellipses to model the motion of orbiting satellites, and you used parametric equations of parabolas to model the path of a kicked soccer ball. In Unit 6, "Polynomial and Rational Functions," you saw that the graph of the inverse power model $y = \frac{1}{x}$ had the shape of a hyperbola with the x-axis as horizontal asymptote and the y-axis as vertical asymptote. In this investigation, you will explore other representations, characterizations, and applications of the conic sections.

1. Recall that the graph of the parametric equations

$$x = r \cos t + h, \qquad y = r \sin t + k, \qquad 0 \le t < 2\pi$$

   is a circle.

   a. What are the center and radius of the circle?

   b. Use the Pythagorean identity $\sin^2 t + \cos^2 t = 1$ to show that the circle can also be represented by the equation $(x - h)^2 + (y - k)^2 = r^2$.

   c. The equation in Part b is called the **standard form of the equation of a circle** with center (h, k) and radius r.

   - Write, in standard form, the equation of a circle with center $(-3, 2)$ and radius 5.

   - Write your equation in expanded form $x^2 + y^2 + ax + by + c = 0$.

   d. The graph of an equation like $x^2 + y^2 + 6x + 10y + 11 = 0$, in which the coefficients of x^2 and y^2 are equal (1, in this case), is a circle. You can use your understanding of the factored form of a perfect square trinomial, $a^2x^2 + 2abx + b^2 = (ax + b)^2$, to rewrite such equations in standard form, revealing the center and radius of the circle.

   - Explain why the following equations are equivalent. In the case of the third equation, also explain why 9 and 25 were chosen to be added.

   $$x^2 + y^2 + 6x + 10y + 11 = 0$$
   $$(x^2 + 6x) + (y^2 + 10y) = -11$$
   $$(x^2 + 6x + 9) + (y^2 + 10y + 25) = -11 + 9 + 25$$
   $$(x + 3)^2 + (y + 5)^2 = 23$$

   - What are the center and radius of the circle?

   - Use the method above, often called *completing the square*, to determine the center and radius of the circle with equation $x^2 + y^2 + 12x - 2y + 21 = 0$. Then sketch a graph of the circle.

   - Why do you think that the method you used is called "completing the square"?

2. You may recall that parametric equations for ellipses with center at the origin are similar to those for circles with center at the origin.

$$x = a \cos t, \qquad y = b \sin t, \qquad 0 \le t < 2\pi$$

a. What are the x-intercepts and y-intercepts of the ellipse?

b. When $a = b$, the equations describe a circle of radius a. When $a > b$, the curve is stretched more along its horizontal line of symmetry. When $a < b$, the curve is stretched more along its vertical line of symmetry.

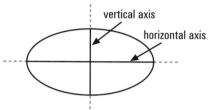

- The portion of the horizontal symmetry line connecting points on the ellipse is called the *horizontal axis of the ellipse*. What is the length of the horizontal axis?

- The portion of the vertical symmetry line connecting the points on the ellipse is called the *vertical axis of the ellipse*. What is the length of the vertical axis?

c. The longer axis of an ellipse is called the *major axis*, and the shorter axis is called the *minor axis*. When will the horizontal axis of an ellipse be the major axis? The minor axis?

d. Sketch the ellipse described by the following parametric equations.

$$x = 2 \cos t, \qquad y = 5 \sin t, \qquad 0 \le t < 2\pi$$

- What is the length of the horizontal axis? Of the vertical axis?

- Sketch and discuss the ellipse described by the following equations.

$$x = 2 \cos t + 5, \qquad y = 5 \sin t + 8, \qquad 0 \le t < 2\pi$$

- Write parametric equations for the ellipse with center (h, k) having horizontal and vertical lines of symmetry, with $2a$ the length of the horizontal axis and $2b$ the length of the vertical axis.

e. Use the Pythagorean identity $\sin^2 t + \cos^2 t = 1$ to show that the ellipse described by the parametric equations you wrote at the end of Part d can also be described by the following equation:

$$\frac{(x - h)^2}{a^2} + \frac{(y - k)^2}{b^2} = 1$$

f. The equation above is called the **standard form of the equation of an ellipse** with center (h, k), $2a$ the length of the horizontal axis, and $2b$ the length of the vertical axis.

- Write in standard form the equation of an ellipse with center $(3, -5)$, horizontal axis of length 12, and vertical axis of length 20.

- Write your equation in expanded form $ax^2 + by^2 + cx + dy + e = 0$.

g. For each equation, determine the nature of its graph by writing the equation in standard form and then sketch the graph.

- $9x^2 + 4y^2 - 36 = 0$
- $3x^2 + 3y^2 - 108 = 0$
- $36x^2 + 9y^2 - 216x = 0$
- $4x^2 + 25y^2 + 24x + 50y - 39 = 0$

3. In your previous work, you saw that graphs of quadratic functions $y = ax^2 + bx + c$, $a \neq 0$, were parabolas. Equivalent forms of a quadratic function rule reveal different information about its graph.

 a. What information can you deduce about the graph of $y = ax^2 + bx + c$ by examining its symbolic form?

 b. What information can you deduce about the graph of $y = a(x - h)^2 + k$ by examining its symbolic form?

 c. You can use the method of completing the square to rewrite a quadratic function in the *vertex form* given in Part b. Rewrite each function rule in vertex form and then describe the graph as completely as possible.

 - $y = x^2 + 6x + 8$
 - $y = 5x^2 - 10x + 9$

 d. Now think back to your work in Unit 2, "Modeling Motion." How did you represent a parabola using parametric equations?

Each of the conics can also be described as a set of points satisfying a specified distance condition. For example, you are familiar with the definition of a *circle* as the set of points in a plane equidistant from a fixed point. Activities 4 and 5 revisit parabolas and ellipses from this perspective. In Activity 6, you will use this *locus-of-points* perspective to derive a general formula for a hyperbola.

4. A *parabola* can be defined as the set of points in a plane such that each point on the parabola is equidistant from a fixed line and a fixed point that is not on the line.

 a. Sketch the set of all points in a plane whose distance from $(0, 0)$ is equal to their distance from the line $y = 3$.

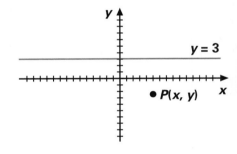

 b. Find an equation of the parabola you drew in Part a. (*Hint:* If $P(x, y)$ is any point in this set, what must be true?)

 c. The point used in defining a parabola is called the *focus* and the line is called the *directrix*. Find the equation of a parabola with focus $F(0, c)$ and directrix $y = -c$.

 - Where is the vertex located?
 - For what values of c is the vertex a maximum point? A minimum point?

 d. Find the equation of a parabola with focus $F(c, 0)$ and directrix $x = -c$.

 - Explain why your equation is not an equation of a function of the form $y = f(x)$.
 - Where is the vertex of this parabola located?
 - How is this parabola similar to, and different from, the parabola in Part c?

Parabolic surfaces (surfaces found by rotating a parabola about its line of symmetry or axis) have useful reflective properties. In the case of parabolic headlight reflectors as on the left below, the bulb is placed at the focus for the high beam and in front of the focus for the low beam. When light is projected from the focus, the light will be reflected in rays parallel to the axis and, in the case of high-beam headlights, parallel to the road surface.

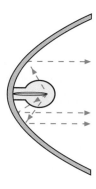

The opposite principle is used in satellite receiving dishes such as that above on the right. In this case, parallel television or radio waves are reflected from the parabolic surface and concentrated at the focal point where the electronic receiver is located.

5. An *ellipse* can be defined as the set of points in a plane for which the sum of the distances from two fixed points, called the *foci* (plural of focus), is a constant.

   a. Suppose the foci of an ellipse are at $F_1(-4, 0)$ and $F_2(4, 0)$, and that for any point $P(x, y)$ on the curve, $PF_1 + PF_2 = 10$. Show that the equation of the ellipse is $9x^2 + 25y^2 = 225$ or $\frac{x^2}{25} + \frac{y^2}{9} = 1$.

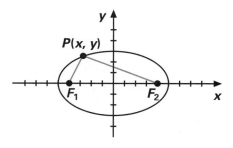

   ■ What is the center of the ellipse?

   ■ What are the lines of symmetry?

   ■ What is the length of the horizontal axis? The vertical axis?

   b. How would your equations and answers to Part a change if the foci were $F_1(0, -4)$ and $F_2(0, 4)$?

c. Now consider a more general case. Suppose the foci of an ellipse are $F_1(-c, 0)$ and $F_2(c, 0)$, and that for any point $P(x, y)$ on the ellipse, the sum of the distances to the foci is $2a$, where $a > c$.

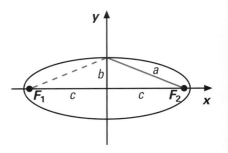

■ Explain why the x-intercepts are $(-a, 0)$ and $(a, 0)$.

■ Explain why the y-intercepts are $(0, -b)$ and $(0, b)$, where $b = \sqrt{a^2 - c^2}$.

■ Use the distance formula and symbolic reasoning to show that the equation of the ellipse can be written in the form
$$\frac{x^2}{a^2} + \frac{y^2}{b^2} = 1.$$

d. Now consider the geometry of the ellipse in Part c.

■ What is the center of the ellipse?

■ What are the lines of symmetry?

■ What is the length of the horizontal axis? The vertical axis?

e. How would you modify the equation in Part c if the line through the foci is parallel to the x-axis and the center is at a point with coordinates (h, k)?

The ellipse, like the parabola, has interesting and useful reflective properties. Any sound or light wave initiated at one focus will be reflected to the other focus. You can experience this phenomenon in "whispering galleries." Statuary Hall in the U.S. Capitol building, shown at the right, is an elliptical chamber. A person whispering at one focus can be easily heard by another person standing at the other focus, even though the person cannot be heard at many places between the foci.

Statuary Hall

This reflective principle of an ellipse is also used in lithotripsy, a procedure for treating kidney stones. The patient is submerged in an elliptical tank of warm water with the kidney stone positioned at one focus. High-energy shock waves generated at the other focus are reflected and concentrated on the stone, pulverizing it.

In the case of an ellipse, the sum of the distances from any point on the curve to the foci is a constant. A different curve, the *hyperbola*, is obtained when the *difference* of the distances from the foci is constant.

6. Suppose the foci of a hyperbola are $F_1(-c, 0)$ and $F_2(c, 0)$. Consider points $P(x, y)$ such that the difference of the distances between $P(x, y)$ and the foci is a constant $2a$, where $a > 0$.

a. If point P is not on the segment connecting the foci, explain why $F_1P + 2c > F_2P$, and therefore $c > a$.

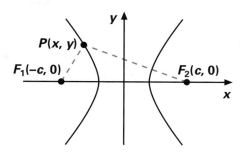

b. Use the distance formula and symbolic reasoning to show that the equation of the hyperbola can be written in the form $\frac{x^2}{a^2} - \frac{y^2}{c^2 - a^2} = 1$.

c. Explain why $c^2 - a^2 > 0$. Since $c^2 - a^2$ is positive, you can let $c^2 - a^2 = b^2$. Then the equation in Part b can be written as $\frac{x^2}{a^2} - \frac{y^2}{b^2} = 1$.

d. Each continuous part of the hyperbola is called a *branch* of the hyperbola. The points on the hyperbola that lie on the segment connecting the foci are called the *vertices* of the hyperbola.

- Explain why the vertices of the hyperbola in Part c are $(-a, 0)$ and $(a, 0)$.

- Explain why that hyperbola has no y-intercepts.

- What are the lines of symmetry?

e. Now consider the hyperbola with equation $9x^2 - 4y^2 = 36$.

- What are the intercepts? The lines of symmetry?

- Explain why the lines $y = \frac{3}{2}x$ and $y = -\frac{3}{2}x$ are asymptotes of this hyperbola.

- Sketch the hyperbola by first sketching its asymptotes.

f. Use asymptotes to help you sketch the graph of $25y^2 - 4x^2 = 100$.

7. Explain how you know that an equation of the form $\frac{(x-h)^2}{a^2} - \frac{(y-k)^2}{b^2} = 1$, or $\frac{(y-k)^2}{a^2} - \frac{(x-h)^2}{b^2} = 1$, is an equation of a hyperbola.

a. What are the lines of symmetry of the graph of each equation?

b. The equations above are called the **standard forms of the equation of a hyperbola** with center (h, k) and constant difference $2a$.

- Write the equation $9x^2 - 16y^2 + 36x + 32y - 124 = 0$ in standard form.

- Describe the graph as completely as possible and make a sketch.

8. In Activity 1, you used the parametric representation of a circle with radius r and center (h, k) to derive the standard form of the equation of a circle. Can you reverse this process and use a different Pythagorean identity to find a parametric representation for a hyperbola with equation of the form $\frac{x^2}{a^2} - \frac{y^2}{b^2} = 1$? Compare your findings with another group and resolve any differences.

Hyperbolas are not as commonly seen in manmade objects as circles, ellipses, and parabolas, but one place you can see them is on the wall of a room behind the lamp shade of a lighted lamp. Also, the LORAN (LOng RAnge Navigation) navigational system is based on locating a vessel at the intersection of two hyperbolas.

Checkpoint

In this investigation, you examined conic sections in terms of their algebraic and geometric representations and properties.

a Describe each of the conics in terms of the intersection of a plane and a double cone.

b Conics whose symmetry lines are parallel to the x-axis and/or y-axis can be expressed in several forms. How can you tell by examining the coefficients of an equation of the form $ax^2 + by^2 + cx + dy + e = 0$, where a and b are not both zero, whether the equation is that of a circle, an ellipse, a parabola, or a hyperbola?

c How can you rewrite an equation like that in Part b so that it is easier to sketch the graph of the curve?

d Write the standard form equation of each of the following conics and explain as completely as possible what that form allows you to conclude about the graph. Assume each conic has the x- and y-axes as symmetry lines.

- Circle
- Ellipse
- Hyperbola

e How would you modify the equations in Part d if the lines of symmetry were parallel to the x- and y-axes?

Be prepared to share your descriptions and thinking with the class.

On Your Own

Identify the conic section represented by each equation and write the equation in standard form. Then use properties of the conic to sketch a graph.

a. $x^2 - 6x + y^2 - 40 = 0$

b. $4x^2 - y^2 - 16 = 0$

c. $3x^2 + 12x + 5y^2 + 30y + 42 = 0$

MORE

Modeling • Organizing • Reflecting • Extending

Modeling

1. The 3.5-mi Ice Cave Trail, marked by a heavy dashed blue line, starts in the upper left-hand portion of the map below, just west of the south end of the dark shaded reservoir. It meets the 8.7-mi Blue Ridge Trail in the center of the map near the marker labeled "107." The intersection is at about 7,200 ft elevation. Contour lines are at 40-foot intervals.

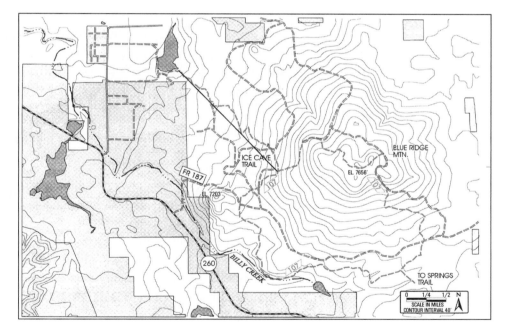

   a. Use the contour map to estimate the elevation of the beginning of the Ice Cave Trail.

   b. Draw a relief map determined by the relief line beginning near the Ice Cave Trail head and ending at the intersection of the two trails.

   c. Based on the maps, estimate the difficulty of hiking the Ice Cave Trail. Explain your reasoning.

2. The diagram below shows the contours of the temperature along one wall of a heated room throughout one winter day, with time indicated as on a 24-hour clock. The room has a heater located at the left-most corner of the wall, and there is one window in the wall. The heater is controlled by a thermostat a few feet from the window. (Adapted from *Multivariable Calculus*, Preliminary Edition by William McCallum, Deborah Hughes-Hallett, Andrew Gleason, et al., New York: Reprinted by permission of John Wiley & Sons, Inc., 1996.)

a. Where is the window? When is it open?

b. Can you explain why the temperature at the window at 5 P.M. is less than at 11 A.M.?

c. When is the heat on?

d. To what temperature do you think the thermostat is set? How do you know?

e. How far is the thermostat from the left-most corner of the wall?

3. The data on the next page were collected at 90-ft intervals by the Pioneer probe of equatorial Venus.

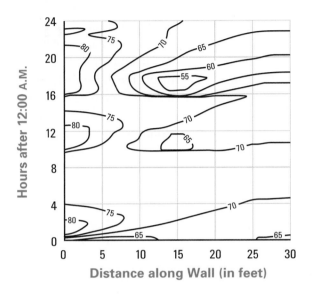

a. Use the data to construct a contour map of the region. Include all values within a 50-ft interval in one contour line to simplify the contour construction.

b. Sketch three parallel vertical cross sections, one at the left edge, one at the middle, and one at the right edge of the region.

c. Are there possible landing sites in this region? If so, identify and justify your site choice(s). If not, explain why there is no appropriate site.

569	554	473	424	449	523	585	623	631	616	587	572	575	565	533	512	504
593	585	526	507	530	570	625	659	663	658	619	587	587	576	525	507	511
593	596	591	606	611	623	658	666	666	661	636	619	624	617	551	527	531
588	608	612	635	660	696	697	675	672	663	652	655	663	649	591	565	565
577	610	623	702	788	809	760	707	703	673	653	657	659	651	613	584	580
560	615	642	735	852	868	835	744	716	696	671	652	638	634	628	611	591
537	615	633	699	869	943	936	795	736	723	676	651	628	628	642	642	621
511	600	628	698	903	1028	998	845	761	747	682	651	633	634	646	661	695
493	556	606	651	807	965	973	856	752	719	682	649	634	633	658	708	753
478	517	582	595	668	824	877	838	740	696	668	630	629	641	701	763	769
476	495	551	567	610	747	834	803	719	686	672	638	631	639	672	729	753
477	488	516	519	529	630	696	671	639	650	648	645	645	627	616	665	724
480	482	495	503	504	538	547	543	566	615	624	637	654	653	674	704	736
479	480	484	498	504	513	527	533	549	600	615	622	632	670	749	771	776
466	468	470	490	504	508	527	542	549	572	587	581	604	648	736	776	794
455	455	457	482	501	503	521	543	540	536	560	570	599	630	699	765	793
441	449	452	466	478	476	483	489	480	477	533	563	574	588	654	748	773

Source: http://nssdc.gsfc.nasa.gov

4. Conic sections can be characterized by *Moiré patterns*—patterns formed by two intersecting sets of concentric circles. Graph paper with intersecting sets of concentric circles is sometimes called *conic graph paper*. In the conic graph paper below, the radius of the smallest circles is 1 unit and the centers of the nonconcentric circles are 10 units apart.

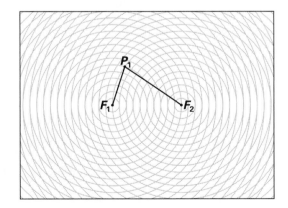

a. Explain why $F_1P_1 + F_2P_1 = 16$ in the diagram above.

b. On a piece of conic graph paper:

- Label the centers of the circles F_1 and F_2.
- Label the point P_1 as in the diagram above.
- Find 20 points P such that $F_1P + F_2P = 16$.
- If you were to connect all points P such that $F_1P + F_2P = 16$, what type of figure would you draw? How do you know? What are the points F_1 and F_2?

c. Use conic graph paper to draw an ellipse with horizontal axis of length 20. Can more than one such ellipse be drawn? Explain.

5. Conic graph paper, as described in Modeling Task 4, is also helpful when drawing hyperbolas. The radius of the smallest circles is 1 unit.

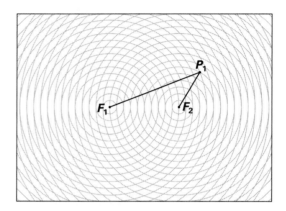

a. On a sheet of conic graph paper, label points F_1, F_2, and P_1 as above.

- Verify that $F_1P_1 = 14$ and $F_2P_1 = 6$.
- Locate all points P such that $F_1P = 14$ and $F_2P = 6$. How many different points satisfy these conditions?

b. Find and mark all points P such that:

- $F_1P = 16$ and $F_2P = 8$
- $F_1P = 8$ and $F_2P = 16$

c. Plot 20 other points P such that $|F_1P - F_2P| = 8$ and then connect your points.

d. What type of figure did you draw in Part c? Justify how you know.

e. Use conic graph paper to draw a hyperbola with vertices that are 4 units apart and that has foci that are 10 units apart.

Organizing

1. In Investigation 2, you were asked to consider a possible rectangular coordinate description of a point in space, namely, an ordered triple of numbers (x, y, z). Rectangular coordinates of points are determined by measuring the directed perpendicular distance from the point to each of three mutually perpendicular planes meeting at a single point called the origin. As shown in the diagram at the right, the intersections of the three sets of pairs of planes can be used to determine a three-dimensional coordinate system. In a three-dimensional rectangular coordinate system, it is customary for the *y*-axis to be horizontal and the *z*-axis to be vertical as shown. The positive *x*-axis appears to come forward out of the paper and creates the appearance of depth or the third dimension.

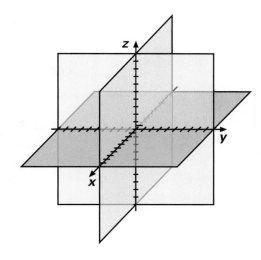

a. How would you describe the *xz*-plane? The *xy*-plane?

b. Into how many regions do these three coordinate planes separate space?

c. Describe the location of all points with

- positive *x*-coordinates;

- negative *z*-coordinates;

- positive *y*- and negative *x*-coordinates.

d. Care is needed in plotting and interpreting points in three-dimensional space. When plotting points, it is helpful to show the "path" to the point as indicated in the diagram below.

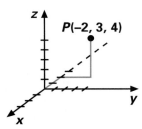

- Plot and draw the paths to points $P(-2, 3, 4)$ and $Q(2, 3, 4)$ on a copy of the three-dimensional coordinate system shown.

- What should appear to be true about the plot of these two points?

e. Plot and label each of the following points on a three-dimensional coordinate system.

- $A(0, 2, 3)$

- $B(3, 2, 4)$

- $C(3, -2, -4)$

- $D(-5, 3, 5)$

- $E(-4, -7, -2)$

f. Describe in general how you would plot a point $A(a, b, c)$.

2. Draw a three-dimensional coordinate system.

   a. Using the origin as a vertex, draw a cube with 4 units on a side with its edges along the positive *x*-, *y*-, and *z*-axes. Find the coordinates of each vertex.

   b. Draw another 4-unit cube with edges along the negative *x*- and *y*-axes and along the positive *z*-axis. Find the coordinates of each vertex.

3. High cholesterol resulting in clogged arteries was once thought to be the major underlying cause of heart attacks. Yet, half of all heart attack victims have cholesterol levels that are normal or even low. New research at Boston's Brigham and Women's Hospital suggests that inflammation, as measured by elevated levels of C-reactive protein, is another important independent trigger.

   Study the three-dimensional bar graph below that shows the relative risk of cardiovascular problems by levels of cholesterol and C-reactive protein.

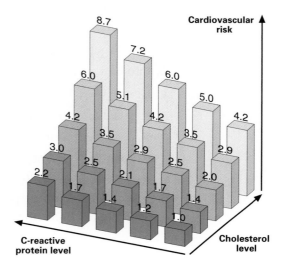

Source: *Circulation*, Vol. 103, No. 13, April 3, 2001.

Explain how each of the following findings is indicated in this plot.

   a. A person with the highest combination of cholesterol and C-reactive protein has almost 9 times the risk as someone with the lowest combination.

   b. A person with C-reactive protein in the top quintile has twice the risk as someone in the lowest quintile with the same cholesterol level.

   c. High cholesterol levels seem to contribute more to the risk of cardiovascular problems than do high C-reactive protein levels.

4. Identify the conic section represented by each equation and write the equation in standard form. Then describe the characteristics of the curve as completely as possible and draw a sketch.

   a. $81x^2 - 36y^2 = 2{,}916$

   b. $5x^2 - 10x - y + 9 = 0$

   c. $3x^2 + 3y^2 - 51 = 0$

   d. $y^2 - x^2 - 8y - 20 = 0$

   e. $4x^2 + y^2 - 8x + 6y + 9 = 0$

5. Write an equation describing each conic section shown below.

a.

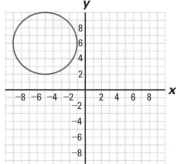

b.

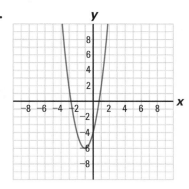

c.

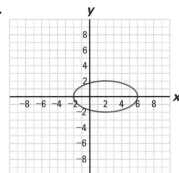

d.

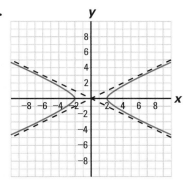

Reflecting

1. Contour maps are informative but do not necessarily convey a complete picture of a region. Here are two contour lines at 40-foot elevation intervals. The horizontal distance between the contours is approximately 300 yards.

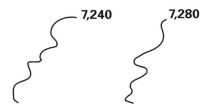

7,240   7,280

   a. Could there be a region between these contours over 7,280 ft? Explain.

   b. Sketch several possible side views of the 300 yards between these contours.

   c. How does the existence of a trail across the region between contours eliminate some possible configurations of the terrain?

2. Find an example of how contour maps are used to communicate the temperature patterns found across the United States and Canada. How is color sometimes used to indicate the higher temperature regions and the lower temperature regions?

3. In Investigation 2, you were introduced to a three-dimensional rectangular coordinate system. It is often helpful to picture a three-dimensional coordinate system in terms of a room you are in. Think of the origin of the coordinate system as a corner at floor level where two walls meet.

a. Describe the x-, y-, and z-axes in this context.

b. What would points with negative coordinates correspond to in this context?

4. Look back at the diagrams of the double cones at the beginning of Investigation 4 (page 527).

a. If possible, describe or sketch a plane that intersects a double cone at a point.

b. If possible, describe or sketch a plane that intersects a double cone in a line.

c. If possible, describe or sketch a plane whose intersection with a double cone is a pair of intersecting lines.

5. A beam of light from a pen light or flashlight forms a cone. When the beam of light is projected onto a flat surface, the image is a conic section. In a darkened room, using a piece of cardboard as a plane and a pen light or flashlight, explore how to place the cardboard relative to the light source to produce each of the four conic sections. Write a brief summary of your findings.

Extending

1. The Global Positioning System (GPS) locates an object's position on the Earth—latitude, longitude, and altitude. The system is very sophisticated. It can identify a location to within a meter or so. Boaters use GPS to locate themselves when they cannot see recognizable landmarks. Satellite navigation systems are now also available as options in automobiles.

Research and write a short report on the capabilities and underlying mathematical features of GPS. Include some detail on at least one other application of GPS.

2. Select one of the following research projects to learn more about the usefulness of contour diagrams.

a. Consult your state's Department of Natural Resources or another agency that has responsibility for lakes and rivers to locate topographical maps of regions in your state. Choose one such map for a region and one for a lake and describe the region and lake on the basis of the data included in the map. Describe also the contour intervals used in these maps.

b. If a local weather forecasting service is nearby, contact it and ask for copies of maps describing the atmospheric pressure patterns over the United States on a particular day. Learn how these pressure maps are used to assist in determining the wind velocity and direction. Make a sketch of the pressure data. Write a report describing how pressure maps are used in the weather forecasting business.

3. You have seen that sketching a hyperbola can be aided by first sketching the asymptotes and then using symmetry of the curve as in the diagram below.

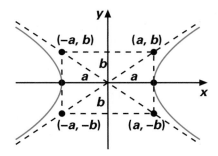

a. If a point with coordinates (p, q) is on the hyperbola with equation $\frac{x^2}{a^2} - \frac{y^2}{b^2} = 1$, use symmetry to identify three other points that are on the graph.

b. Provide an argument that the asymptotes of the hyperbola with equation $\frac{x^2}{a^2} - \frac{y^2}{b^2} = 1$ are $y = \pm\frac{b}{a}x$.

c. Provide an argument that the asymptotes of the hyperbola with equation $\frac{y^2}{a^2} - \frac{x^2}{b^2} = 1$ are $y = \pm\frac{a}{b}x$.

4. Use symbolic reasoning to rewrite the parametric equations

$$x = 4 \sec t, \qquad y = 5 \tan t, \qquad 0 \le t < 2\pi$$

as a single equation in variables x and y. Describe as completely as possible the curve described by the equation. Then use a graphing calculator or computer software to confirm your reasoning.

5. Let C be a circle with center at the origin and radius r. Find an equation for the set of all points $P(x, y)$ that are equally distant from the circle C and the x-axis.

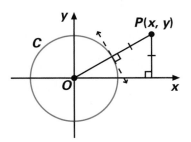

1. If $\frac{1}{5}$ of a class of 40 students is absent, what percent are present?

   (a) 8% (b) 80% (c) 20% (d) 0.20% (e) 92%

2. Simplify: $\dfrac{\frac{25x^2-1}{x-7}}{\frac{5x+1}{7-x}}$

   (a) $1-5x$ (b) $-5x$ (c) $\dfrac{5}{49-x^2}$ (d) $5x-1$ (e) $5x$

3. How many solutions does this system have?

   $$-4x + 3y = 4$$
   $$4x - 3y = -5$$

   (a) none (b) an infinite number (c) two (d) one (e) three

4. Given the graph of $y = f(x)$, which of the following could be a portion of the graph of $y = f(x + 2)$?

   $y = f(x)$

   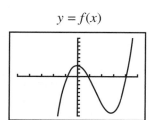

   (a) 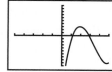 (b)  (c)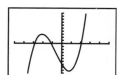

   (d) 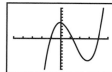 (e)

5. Simplify: $(6 + 5i)(9 - 4i)$, where $i = \sqrt{-1}$.

    (a) $30 + 2i$ **(b)** $30 - 21i$ **(c)** $74 + 4i$ **(d)** $74 + 21i$ **(e)** $35 + 21i$

6. Solve: $17 - 8(x - 4) - 3(5x + 8) = -15 + 3(x - 4)$.

    (a) $x = 2$ **(b)** $x = 10$ **(c)** $x = \frac{1}{2}$ **(d)** $x = -\frac{17}{9}$ **(e)** $x = -5$

7. Which of the following could be a portion of the graph of $y = \dfrac{x + 2}{(x + 3)(x - 5)}$?

  (a) 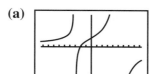   **(b)**    **(c)**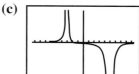

  (d) 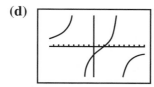   **(e)**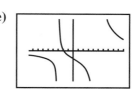

8. If $\sin^2 \theta + 4 = x - \cos^2 \theta$, then x is

    (a) 3 **(b)** 5 **(c)** $\sin^2 \theta - \cos^2 \theta + 4$ **(d)** $\dfrac{\sin^2 \theta + 4}{\cos^2 \theta}$ **(e)** 4

9. If $\log x + \log x = \log 36$, then

    (a) $x = 18$ **(b)** $x = 6$ or $x = -6$ **(c)** $x = 72$ **(d)** $x = 3.11$ **(e)** $x = 6$

10. Simplify: $\sqrt[3]{24x^6} + \sqrt[3]{250x^6} - \sqrt[3]{8x^3}$.

    (a) $7x^2\sqrt[3]{5} - 2x$ **(b)** $2x^2\sqrt[3]{3} + 5x^2\sqrt[3]{2} - 2x$ **(c)** $x^3\sqrt[3]{3} + 5x^2\sqrt[3]{2}$

    (d) $7x^2\sqrt[3]{6} - 2x$ **(e)** $7x^2(\sqrt[3]{3} + \sqrt[3]{2} - 2)$

Equations for Surfaces

In the previous lesson, you represented familiar space-shapes using perspective sketches, cross sections, and contour diagrams. You also explored the use of three-dimensional coordinates. Computer models of space-shapes like the one at the right are based on coordinate representations. The images are often produced from information about cross sections derived from the equations of the surfaces of the shapes.

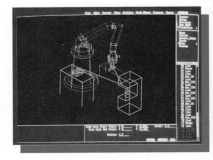

Courtesy of Computervision

You can use your understanding of coordinate representations of two-dimensional shapes to guide your thinking about coordinate representations of three-dimensional shapes.

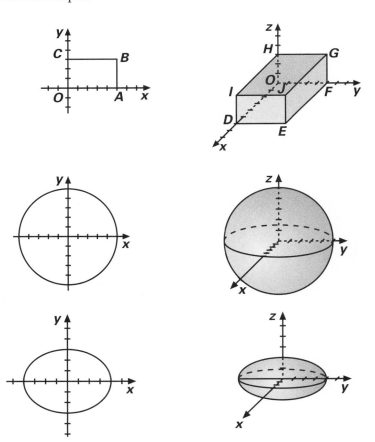

Use the diagrams on the previous page to help you think about possible similarities and differences between two-dimensional and three-dimensional shapes and their coordinate representations. The scales on the axes are 1 unit.

a What are the coordinates of point *B*, a vertex of the rectangle? What are the coordinates of point *J*, a vertex of the rectangular prism? How was your reasoning similar in each case?

b What are the coordinates of the point symmetric to point *B* with respect to the *y*-axis? With respect to the *x*-axis? What do you think are the coordinates of the point symmetric to point *J* with respect to the *xz*-plane? With respect to the *xy*-plane? Explain your reasoning.

c What is true about all points on the line determined by points *B* and *C*? What is the equation of the line? What is true about all points on the plane determined by points *I*, *D*, and *E*? What do you think is the equation of the plane?

d How would you determine the length of $\overline{AC}$? How might you determine the length of $\overline{DG}$?

e What is the equation of the circle? Of the ellipse? What equation do you think would describe the *sphere*? The *ellipsoid*?

INVESTIGATION 1 ▶ Relations Among Points in Three-Dimensional Space

In the MORE set in Lesson 1, you explored how to represent points in space with *x*-, *y*-, and *z*-coordinates. The *x*-axis and the *y*-axis determine the *xy-plane* as shown below. Similarly, the *x*-axis and the *z*-axis determine the *xz-plane*, and the *y*-axis and *z*-axis determine the *yz-plane*.

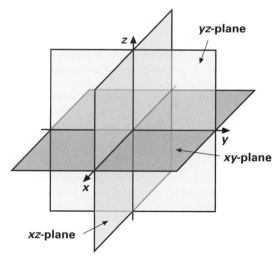

1. You can determine the equations of the *coordinate planes* and planes parallel to them by reasoning from analogous cases in a two-dimensional coordinate system.

   a. Think about the form of equations of horizontal and vertical lines in a two-dimensional (x, y) coordinate system.

   ■ What is the equation of the x-axis? Of the y-axis? Why do these equations make sense?

   ■ What are the equations of the lines 5 units away from and parallel to the x-axis?

   ■ What are the equations of the lines 10 units away from and parallel to the y-axis?

   b. Now think about the form of equations of horizontal planes in a three-dimensional (x, y, z) coordinate system.

   ■ Which coordinate plane is represented by the equation $z = 0$? Why does this make sense?

   ■ What is the equation of the plane 4 units above and parallel to the xy-plane?

   ■ What is the equation of the plane 6 units below and parallel to the xy-plane?

   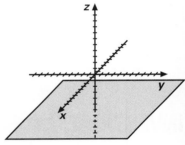

   ■ Shown at the right is a sketch of one of the two planes described above. Reproduce this diagram and then sketch the other plane. Label both planes.

   c. What is the equation for the xz-plane? What are the equations of the planes 5 units away from and parallel to the xz-plane? Sketch and label all three planes on the same coordinate system.

   d. What is the equation for the yz-plane? What are the equations of the planes 12 units away from and parallel to the yz-plane? Sketch and label those two planes on the same coordinate system.

2. In a previous course, you derived a formula for computing distances between pairs of points in a coordinate plane. If $P(x_1, y_1)$ and $Q(x_2, y_2)$ are points in a coordinate plane, then

$$PQ = \sqrt{(x_1 - x_2)^2 + (y_1 - y_2)^2}.$$

You used this distance formula to help derive equations for the conic sections in Lesson 1.

   In the "Think About This Situation" at the beginning of this lesson, you considered how you might compute the distance between points in three-dimensional space. Compare your ideas with the following approach (shown on the next page) suggested by one class in Ann Arbor, Michigan. Their approach is based on the rectangular prism at the beginning of this lesson.

a. To find the length of $\overline{DG}$, the students suggested drawing $\overline{DF}$ and then using the Pythagorean theorem twice.

- Why is $\triangle DEF$ a right triangle? What is the length of $\overline{DF}$?

- Why is $\triangle DFG$ a right triangle? What is the length of $\overline{DG}$?

- What is the length of $\overline{OJ}$ in the diagram at the right?

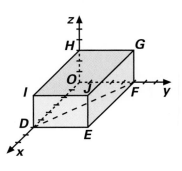

b. Use the diagram below to derive a formula for finding the distance PQ if $P(x_1, y_1, z_1)$ and $Q(x_2, y_2, z_2)$ are points in three-space and the figure shown is a rectangular prism.

- What are the coordinates of point S? Of point R?

- Find PQ.

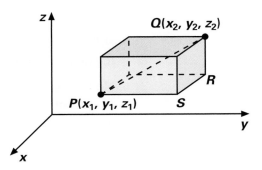

c. Use the distance formula you developed in Part b to find the distance between the points $A(1, 2, 3)$ and $B(6, -2, -7)$.

3. Recall that a circle is the set of points in a plane at a given distance from a fixed point, its center. In three-dimensional space, the set of points at a given distance from a fixed point is a **sphere**. Like a circle, a sphere is determined by its center (the fixed point) and its radius (the given distance).

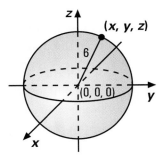

a. Find the equation for a sphere with center at the origin and radius 6.

b. What are the x-, y-, and z-intercepts of the sphere in Part a? Do the coordinates of these points satisfy your equation?

c. Using the diagram below, derive an equation for a sphere with center at the origin and radius *r*.

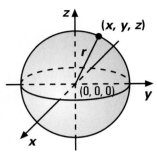

d. The equation derived in Part c represents the surface of the sphere. How would you represent the solid ball enclosed by the sphere?

4. The distance formula in three-dimensional space is a generalization of the distance formula in two-space. Now, investigate how to generalize the formula for the midpoint of a segment in a plane to that of a segment in three-space.

a. If $P(x_1, y_1)$ and $Q(x_2, y_2)$ are points in a plane, what are the coordinates of the midpoint of $\overline{PQ}$?

b. Now consider the points $A(5, 7, 12)$ and $B(2, -4, 4)$. Use your knowledge of computing midpoints in a plane to calculate the midpoint of $\overline{BC}$. Then use your result to calculate the midpoint of $\overline{AB}$. Use the distance formula to verify that the point you found is the midpoint.

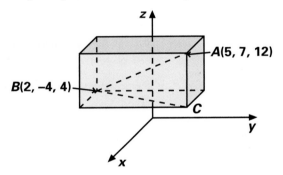

c. Use the diagram in Activity 2 Part b to derive a general midpoint formula in three-space. Compare your formula for the midpoint of $\overline{PQ}$ to those of other groups. Resolve any differences.

5. Now pull together the ideas you developed in this investigation. Consider the points $S(-3, 1, 5)$ and $T(2, 4, -3)$ in three-space.

a. Find the equation of the plane containing point S and parallel to the xy-plane.

b. Find the equation of the plane containing point T and parallel to the yz-plane.

c. Find ST.

d. Find the coordinates of the midpoint of $\overline{ST}$.

e. Find the equation of the sphere with center at the origin and containing point T.

In this investigation, you extended important ideas involving coordinates in two dimensions to coordinates in three dimensions.

ⓐ In three-dimensional space, how are the equations and graphs of planes parallel to a coordinate plane similar to, and different from, the equations and graphs of lines parallel to a coordinate axis in two-dimensional space?

ⓑ Describe how the formula for the distance between two points in a three-dimensional coordinate system is similar to, and different from, the distance formula for points in a two-dimensional coordinate system.

ⓒ Describe how to find the coordinates of the midpoint of a segment in three-space.

ⓓ How is the equation of a sphere with center at the origin similar to, and different from, the equation of a circle with center at the origin?

Be prepared to share your ideas with the entire class

On Your Own

In the diagram below, vertex T of the rectangular prism has coordinates $(12, 3, 6)$ and O is the origin.

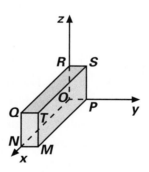

a. Determine the coordinates of the remaining vertices.

b. Write equations for the planes that contain the faces of the prism.

c. Use right triangles to explain why $(PQ)^2 = (PM)^2 + (MN)^2 + (NQ)^2$.

d. Calculate PQ using the equation in Part c and then using the distance formula.

e. Find the midpoints of $\overline{OT}$ and $\overline{RM}$. What can you conclude?

f. Does the sphere with equation $x^2 + y^2 + z^2 = 200$ enclose the prism? Explain.

INVESTIGATION 2 The Graph of $Ax + By + Cz = D$

M.C. Escher's "Three Spheres I"

In two-space, there are two dimensions, two coordinates locate a point, and equations in two variables can be used to specify lines or curves. For example, the graph of $2x - 3y = 12$ is a line and the graph of $\frac{x^2}{25} - \frac{y^2}{10} = 1$ is a hyperbola.

In three-space, there are three dimensions, three coordinates locate a point, and equations in three variables can be used to specify surfaces. The surface may be flat, as in the case of a plane, or curved, as in the case of a sphere. The challenge, often, is to figure out what the surface looks like given its equation and to make a sketch of it, or to interpret the reasonableness of computer- or calculator-produced graphs.

1. The equation $Ax + By + Cz = D$ in three-space is analogous to the equation $Ax + By = C$ in two-space. Based on this analogy, what do you think the graph of $Ax + By + Cz = D$ could be? Explain your reasoning.

2. Now consider a specific instance of the equation $Ax + By + Cz = D$, namely $2x + 4y + 3z = 12$.

   a. Reasoning by analogy from your previous work in a two-dimensional coordinate system, what are the points at which the surface intersects the x-, y-, and z-axes?

   b. Cross sections of the surface formed by the intersection of the coordinate planes or planes parallel to a coordinate plane with the surface can provide useful information about the nature of the surface. The first ones to check are the cross sections made by the coordinate planes. These cross sections are called **traces**.

   - Explain why you can find the trace of the surface in the xy-plane by setting $z = 0$ and examining the resulting equation. Find the equation of the xy-trace. What is the shape of the trace?

   - Find the equation and describe the shape of each of the other two traces.

   - Based on your information about the traces of the graph of $2x + 4y + 3z = 12$, what do you think the surface is? Why?

   c. Additional information about the surface can be obtained by examining the equations of cross sections found by setting x, y, or z to a constant value.

   - Explain how setting $z = 8$ generates a cross section of $2x + 4y + 3z = 12$. That is, how do you know the cross section is in a plane parallel to one of the coordinate planes? Describe the location of the cross section.

   - Find the equation of the cross section determined by setting $z = 8$. What is the shape of this cross section? How is this cross section related to the xy-trace?

   - Describe the cross section formed by setting $x = 5$; by setting $y = 2$.

   d. Sketch the surface $2x + 4y + 3z = 12$ on a three-dimensional coordinate system. Describe the surface.

3. Illustrate how you could use the points where the graph of $2x + 3y + 2z = 6$ intersects the x-, y-, and z-axes to quickly sketch the surface on a three-dimensional coordinate system. Explain why your method works.

4. Sketch the graph of $3x + 5z = 15$ on a three-dimensional coordinate system. Describe the surface.

5. Once again consider the general equation $Ax + By + Cz = D$. Describe the graph of this equation for each set of conditions below. Also identify the traces, the intercepts (points where graph intersects the x-, y-, and z-axes), and the cross section at $z = 4$.

   a. A, B, C, and D are all nonzero.

   b. A, B, and D are nonzero, but $C = 0$.

   c. $A = B = 0$, while C and D are nonzero.

6. Now consider how you can work backward from information about traces to the equation of a surface. Suppose a plane has traces with equations $x + y = 8$, $x + 4z = 8$, and $y + 4z = 8$.

   a. Sketch the plane.

   b. Find an equation for the plane.

7. As you have seen, equations of planes in three-space are similar to equations of lines in two-space. An important characteristic of lines is that they are straight; they have constant slope. Planes have an analogous characteristic: they are flat. But what makes them flat? Consider the portion of the graph of $2x + 3y + 6z = 6$ shown at the right.

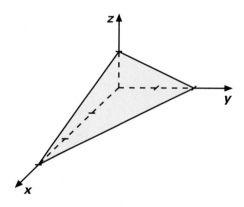

   a. How can you quickly check the reasonableness of this graph?

   b. Now imagine starting at the point $(0, 0, 1)$ and walking around on this plane. What can you say about the "slope" of your walk if you stay directly above the x-axis? The y-axis?

   c. Suppose you walk on the plane along a path with constant y-coordinate. What can you say about the "slope" of your path? Explain your reasoning.

   d. Suppose you walk on the plane along a path with constant x-coordinate. What can you say about the "slope" of your path in this case? Why does this make sense?

   e. What can you say about the "slope" of walks on the plane along paths with the z-coordinate constant?

8. In your previous work in two-space, you saw that two lines could intersect in 0, 1, or infinitely many points. Consider the analogous situation in three-space in the case of the graphs of $2x + 3y + 4z = 12$ and $3x + 2y - z = 6$.

   a. Sketch the planes and describe their intersection.

   b. What are the possible ways the graphs of two distinct equations of the form $Ax + By + Cz = D$ can intersect?

   c. Under what conditions would the graphs of $A_1x + B_1y + C_1z = D_1$ and $A_2x + B_2y + C_2z = D_2$ be parallel?

   d. Is it possible for the graphs of three equations of the form $Ax + By + Cz = D$ to intersect in 0 points? One point? Infinitely many points? Explain.

Checkpoint

The equation $Ax + By + Cz = D$, where not all A, B, and C are zero, is a first-degree (or linear) equation in x, y, and z.

ⓐ What is the graph of an equation of this form?

ⓑ How can you quickly sketch the graph?

ⓒ What is the nature of each cross section of the graph of $Ax + By + Cz = D$?

ⓓ How are the equations of planes parallel to coordinate planes special cases of $Ax + By + Cz = D$?

Be prepared to explain your conclusions to the entire class.

▶ **On Your Own**

Sketch the graph of each equation on a three-dimensional coordinate system.

a. $2x - y + 3z = 6$　　b. $x + y - 2z = 30$　　c. $x - 4y = 8$

INVESTIGATION 3 Surfaces Defined by Nonlinear Equations

In Investigation 2, you found that your understanding of linear equations made sketching planes defined by linear equations of the form $Ax + By + Cz = D$ much easier. Similarly, to sketch surfaces defined by nonlinear equations, you can draw on your understanding of curves, particularly the conics, in a coordinate plane. In addition to locating the points at which the surface intersects the x-, y-, and z-axes and finding cross sections, it is often helpful to draw on ideas of symmetry in three-dimensional space.

1. Re-examine the ellipse and corresponding surface, called an *ellipsoid*, from the beginning of this lesson.

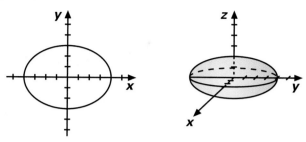

   a. Is the ellipse symmetric with respect to the *x*-axis? With respect to the *y*-axis? Explain your reasoning.

   b. If a point $A(a, b)$ is on the ellipse, name the coordinates of at least two other points on the ellipse.

   c. The ellipsoid above is *symmetric with respect to the xz-plane*. If a point $A(a, b, c)$ is on the ellipsoid, then the point $A'(a, -b, c)$ is also on the ellipsoid.

   ■ Is the ellipsoid *symmetric with respect to the xy-plane*? If so, what are the coordinates of the point that is symmetric to point *A* with respect to the *xy*-plane?

   ■ Is the ellipsoid *symmetric with respect to the yz-plane*? If so, what are the coordinates of the point that is symmetric to point *A* with respect to the *yz*-plane?

2. Now consider the algebraic representations for the ellipse and ellipsoid above.

   a. Explain why an equation for the ellipse is $\frac{x^2}{16} + \frac{y^2}{9} = 1$.

   b. Note that when you substitute $-x$ for x in the equation in Part a, you get $\frac{(-x)^2}{16} + \frac{y^2}{9} = 1$ or $\frac{x^2}{16} + \frac{y^2}{9} = 1$, which is the same as the original equation.

   ■ Why can you use this fact to conclude that the graph of the equation is symmetric with respect to the *y*-axis?

   ■ How could you determine that the graph of $\frac{x^2}{16} + \frac{y^2}{9} = 1$ is symmetric with respect to the *x*-axis by only reasoning with the symbolic form?

   c. In the "Think About This Situation" at the beginning of this lesson, you may have conjectured that an equation for the ellipsoid is $\frac{x^2}{9} + \frac{y^2}{16} + \frac{z^2}{4} = 1$. Explain why this graph-equation match makes sense in terms of each of the following:

   i. *x*-, *y*-, and *z*-intercepts

   ii. cross sections determined by the coordinate planes and planes parallel to the coordinate planes

   ■ What are the equation and shape of the cross section of the graph of $\frac{x^2}{9} + \frac{y^2}{16} + \frac{z^2}{4} = 1$ determined by the *yz*-plane? The *xz*-plane? The *xy*-plane?

   ■ What are the equation and shape of the cross section determined by setting $z = 2$? By setting $y = 2$? By setting $x = 1$?

symmetry

- What symmetry of the graph of $\frac{x^2}{9} + \frac{y^2}{16} + \frac{z^2}{4} = 1$ is implied by the fact that $\frac{x^2}{9} + \frac{y^2}{16} + \frac{(-z)^2}{4} = \frac{x^2}{9} + \frac{y^2}{16} + \frac{z^2}{4} = 1$?
- How could you test for symmetry of the graph of $\frac{x^2}{9} + \frac{y^2}{16} + \frac{z^2}{4} = 1$ with respect to the *xz*-plane by reasoning with the equation itself? With respect to the *yz*-plane?

3. Use analysis of intercepts, cross sections, and symmetry to help you match each equation with one of the following surfaces.

   a. $x^2 + y^2 - z^2 = 0$

   b. $x^2 + 4y^2 - z^2 = 4$

   c. $x^2 + 2y^2 + 3z = 6$

   d. $x^2 + y^2 - z = 0$

i.

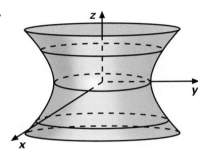

ii.

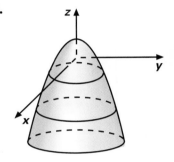

iii.

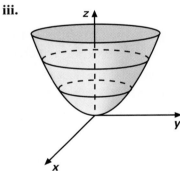

iv.

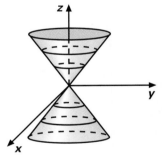

4. Use analysis of intercepts, cross sections, and symmetry to help you visualize and sketch surfaces given by each of the following equations. Describe each surface. Compare your surfaces and descriptions to those of other groups and resolve any differences.

   a. $x^2 - y + z^2 = 1$

   b. $z^2 - x^2 - y^2 = 16$

Graphing surfaces in three-space can be time consuming and difficult. Three-dimensional graphs are more easily produced using computer software or a calculator with three-dimensional graphing capability. Another advantage of these technologies is that they permit you to view the surface from different angles.

5. Three-dimensional graphing software and calculators graph equations of the form $z = f(x, y)$.

   a. The TI-92 calculator display below shows the graph of a portion of the surface in Activity 4 Part a. Explain differences between this graph and the one you sketched.

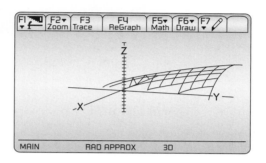

   b. Being able to judge the reasonableness of computer- or calculator-produced graphs is as important when working in a three-dimensional coordinate system as when working in a two-dimensional coordinate system. Explain why it is reasonable that the plot shown in the screen below is that of $x^2 - y^2 + z = 0$.

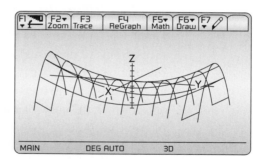

   c. (*Optional*) Explore the use of computer software or a graphing calculator to graph each of the following equations. Describe the general shape of each surface. Experiment with viewing the surface from different angles.

   ■ $z = \dfrac{x^2y - y^2x}{400}$

   ■ $z = 2x - 3y + 3$

   ■ $z = \sqrt{25 - x^2 - y^2}$

   ■ $z = \dfrac{x^2y^2 - y^2x}{400}$

In this investigation, you examined effective strategies for graphing nonlinear equations involving three variables. Suppose you are given the equation of a surface in variables *x*, *y*, and *z*.

ⓐ How would you find the *x*-, *y*-, and *z*-intercepts of the surface?

ⓑ How would you find the traces? Cross sections? Explain how traces and cross sections are similar and different.

ⓒ Explain how you could use the equation to determine if the surface has any symmetries.

Be prepared to explain your procedures to the entire class.

On Your Own

Sketch and describe each of the following surfaces.

a. $(x - 2)^2 + y^2 + (z - 1)^2 = 25$

b. $x^2 + y^2 + z = 4$

INVESTIGATION 4 Surfaces of Revolution and Cylindrical Surfaces

Graphs of equations in three variables are surfaces. Some of these surfaces can be generated by rotating (or revolving) a curve about a line, sweeping out a **surface of revolution**. The line about which the curve is rotated is called the **axis of rotation**. A table leg or lamp base turned on a lathe has a surface of revolution. A potter using a potting wheel makes surfaces of revolution.

Some common surfaces can be thought of as surfaces of revolution.

1. Sketch a graph of $x^2 + y^2 = 25$ in the *xy*-plane of a three-dimensional coordinate system.

 a. Imagine rotating the circle about the *y*-axis. What kind of surface is formed?

 b. Would you get the same surface if the circle was rotated about the *x*-axis? Explain your reasoning.

 c. Write the equation for this surface of revolution.

2. Next consider the segment determined by the points $P(0, 5, 1)$ and $Q(0, 5, 6)$.

   a. Imagine rotating the segment about the z-axis. What kind of surface is formed? Draw a sketch of the surface.

   b. If $A(x, y, z)$ is a point on the surface of revolution, what conditions must be satisfied by x, y, and z?

   c. Develop an equation for this surface of revolution.

3. Describe, and illustrate with a sketch, how a cone with vertex at the origin can be generated as a surface of revolution.

4. Reproduced at the right is the paraboloid from Lesson 1. Write the equation of a curve in the yz-plane that would produce a similar surface when rotated about an axis. What is the axis of rotation?

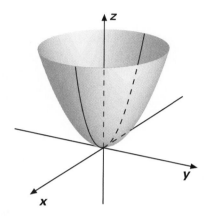

In the next four activities, you will examine surfaces formed by other familiar curves rotated about the x-, y-, or z-axis.

5. Consider the curve $y - x^2 = 0$ in the xy-plane. Imagine the curve revolving about the y-axis to generate a surface.

   a. Sketch the traces of this surface.

   b. Describe the cross sections parallel to the xz-plane.

   c. Sketch the surface.

6. Next consider the curve $z - y^4 = 0$ in the yz-plane.

   a. Sketch the surface generated by rotating this curve about the z-axis.

   b. Why are the cross sections formed by planes parallel to the xy-plane circles?

   c. To find an equation for this surface, choose a point $P(x, y, z)$ on the surface. Then P is on a circular cross section with radius r.

      ■ Explain why $x^2 + y^2 = r^2$, where r is a function of z.

      ■ Explain why for a point $A(0, y, z)$, $|y| = r$.

      ■ Explain why $z = r^4$.

      ■ Explain why $z = (x^2 + y^2)^2$.

7. Use algebraic and geometric reasoning to develop a possible equation for the paraboloid in Activity 4. Based on your work, what can you conclude about the equation for the surface in Activity 5?

8. Now consider a line in the yz-plane that makes an angle of θ, $0° < \theta < 90°$, with the positive z-axis. Generate a surface by rotating the line about the z-axis.

   a. Make a sketch of the line (you choose θ) and the surface generated. Describe the surface.

   b. Sketch the cross section of the intersection of the surface and a plane perpendicular to the z-axis, other than the xy-plane.

   c. To find the equation of this surface, you can use reasoning similar to that used in Activities 6 and 7. Let $P(x, y, z)$ be any point on the surface.

     ■ Explain why $x^2 + y^2 = r^2$, where r is a function of z.

     ■ Explain why $r = z \tan \theta$.

     ■ Explain why $x^2 + y^2 = kz^2$, where $k = \tan^2 \theta$.

   d. What are the equations of the traces of this surface?

   e. Use the equation of the surface to determine the shapes of cross sections parallel to the yz- and xz-planes.

In Activity 2, you formed a cylinder by rotating a segment perpendicular to the xy-plane about the z-axis. More generally, a **cylindrical surface** can be generated by moving a line along the path of a plane curve. The cross sections do not need to be circles as in pipes and cans. They can have a variety of shapes. In addition, cylindrical surfaces need not have a closed cross section such as a circle, hexagon, or ellipse. The cross section could be open like a parabola or a hyperbola.

   Below are sketches of two cylindrical surfaces, one closed and the other open.

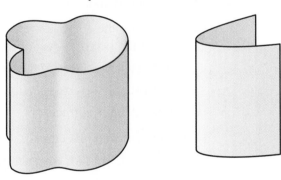

9. Sketch a graph of $y - x^2 = 0$ in the xy-plane and then consider the following actions:

   ■ Consider a line parallel to the z-axis (perpendicular to xy-plane) intersecting the graph.

   ■ Imagine sweeping out a surface by moving the line along the curve $y - x^2 = 0$, keeping the line parallel to the z-axis at all times.

   The surface is a *parabolic cylindrical surface*. Its equation is $y - x^2 = 0$ since z takes on all values for every xy-pair satisfying $y - x^2 = 0$.

   a. Describe the cross sections parallel to the xy-plane.

b. Describe the cross sections parallel to the other coordinate planes.

c. Sketch the parabolic cylindrical surface on a three-dimensional coordinate system.

10. Sketch graphs of the following surfaces on a three-dimensional coordinate system. If possible, check your graphs using three-dimensional graphing software or a calculator with that capability.

   a. The logarithmic cylindrical surface with equation $y = \log z$

   b. The elliptic cylindrical surface with equation $\frac{x^2}{4} + \frac{z^2}{9} = 1$

Checkpoint

In this investigation, you examined how to generate surfaces in three-dimensional space.

ⓐ Describe two ways that plane curves can be used to generate surfaces.

ⓑ How can you use cross sections parallel to a coordinate plane to identify a surface of revolution?

ⓒ Describe a general procedure to develop an equation for a surface of revolution.

ⓓ What kind of surface is defined by an equation with only two variables? What is the effect of the omitted variable?

Be prepared to share your descriptions and thinking with the class.

▶On Your Own

Consider surfaces that can be generated using an ellipse.

a. Sketch the graph of $\frac{x^2}{4} + \frac{y^2}{9} = 1$ in the *xy*-plane of a three-dimensional coordinate system.

b. Sketch the surface of revolution generated by rotating $\frac{x^2}{4} + \frac{y^2}{9} = 1$ about the *y*-axis. Then find the equation of the surface of revolution.

c. Sketch the cylindrical surface with equation $\frac{x^2}{4} + \frac{y^2}{9} = 1$.

d. Compare the equations of the surfaces in Parts b and c and summarize the information revealed by the two different forms.

Modeling • Organizing • Reflecting • Extending

Modeling

1. In this lesson, you saw that geometric ideas such as distance, shape, and symmetry in a two-dimensional coordinate model had analogous representations in a three-dimensional coordinate model. Complete a table like the one below, which summarizes some of the key features of a three-dimensional coordinate model.

Geometric Idea	Two-Dimensional Coordinate Model	Three-Dimensional Coordinate Model
Point	Ordered pair (x, y) of real numbers	
Plane	All possible ordered pairs (x, y) of real numbers	
Distance	For points $A(x_1, y_1)$ and $B(x_2, y_2)$, $AB = \sqrt{(x_1 - x_2)^2 + (y_1 - y_2)^2}$.	
Midpoint	For points $A(x_1, y_1)$ and $B(x_2, y_2)$, midpoint of $\overline{AB}$ is $M\left(\frac{x_1 + x_2}{2}, \frac{y_1 + y_2}{2}\right)$.	
Reflection Symmetry	Across the x-axis $(x, y) \rightarrow (x, -y)$ Across the y-axis $(x, y) \rightarrow (-x, y)$	Across the xz-plane $(x, y, z) \rightarrow$? Across the yz-plane Across the xy-plane
Half-Turn Symmetry	About $(0, 0)$ $(x, y) \rightarrow (-x, -y)$	About the z-axis About the y-axis About the x-axis
Shapes	Circle $x^2 + y^2 = r^2$ Ellipse $\frac{x^2}{a^2} + \frac{y^2}{b^2} = 1$ Parabola $y = ax^2$	Sphere Ellipsoid Paraboloid

2. Triangle *PQR* has vertices *P*(1, 2, 3), *Q*(5, 4, 1), and *R*(−1, 6, 5).

   a. Draw △*PQR* in a three-dimensional coordinate system.

   b. What kind of triangle is △*PQR*?

   c. Find, plot, and label the coordinates of the midpoints of each side.

   d. Is △*PQR* similar to the triangle with the midpoints as vertices? Justify your response.

3. The table below gives the ratings of portable CD players by the editors of a magazine. A rating of 5 is the highest rating in each category. The headphones rating reflects the quality of the sound through the headphones. The error correction rating indicates how well the CD player plays smudged or scratched disks. The bumps rating indicates performance during a range of shakes in the test lab.

	Headphones	Error Correction	Bumps
Brand A	4	5	4
Brand B	3	5	5
Brand C	3	3	5
Brand D	2	5	5
Brand E	4	2	3
Brand F	3	4	3

   a. The three-dimensional plot below provides a graphical model for the ratings shown in the above table. What does each axis represent?

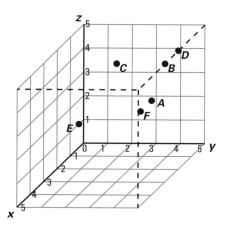

b. Rank the CD players by adding the three ratings.

c. Explain how the distance formula might be used to rank the CD players. Then rank them using your described method. Are the rankings the same or different than the ones in Part b? Which method would you recommend and under what conditions?

d. Suppose you use your CD player primarily while sitting at a desk. In this case, its performance during a range of shakes is not as important to you and the quality of sound is very important. How might you modify your method from Part b to take this into consideration?

e. The magazine also included ratings for the time that it took to move from track to track. How might you modify the distance formula to take this fourth rating into account?

4. A surface is generated by rotating about the z-axis a line that makes a $45°$ angle with the z-axis and contains the origin.

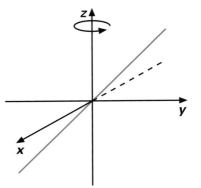

a. Describe the traces.

b. Describe the cross sections parallel to each of the three coordinate planes.

c. Make a sketch of the surface.

d. Derive an equation for this surface.

5. Sketch surfaces given by the following equations. If possible, verify your sketches using three-dimensional graphing software or a calculator with that capability.

a. $x^2 + y^2 = 9$

b. $x^2 + y^2 + z^2 = 9$

c. $x^2 + 2y^2 + 3z^2 = 12$

d. $\frac{x^2}{4} + \frac{y^2}{9} - \frac{z^2}{16} = 1$

e. $\frac{y^2}{4} - \frac{x^2}{9} - \frac{z^2}{9} = 1$

f. $4x^2 + 9y^2 - 36z = 0$

g. $2x - 3y + z - 6 = 0$

Organizing

1. For each of the following equations, describe the surface and identify the x-, y-, and z-intercepts, coordinate plane symmetries, and traces.

a. $x^2 + y^2 - 2z = 0$

b. $x^2 + y^2 + z^2 = 9$

c. $\frac{x^2}{6} + \frac{y^2}{4} + \frac{z^2}{3} = 1$

d. $x^2 + y^2 - z^2 = 0$

e. $x^2 + y^2 - z^2 = -1$

f. $6x - 5y + 10z = 30$

2. Spheres are the three-dimensional analog of circles in a plane. In this lesson, you discovered that the equation of a sphere with center at the origin $(0, 0, 0)$ and radius r is similar to that for a circle with center at the origin $(0, 0)$ and radius r. Use reasoning by analogy to help you complete the following tasks.

   a. Describe as completely as possible the surface with equation
   $(x - 3)^2 + (y - 2)^2 + (z + 5)^2 = 36$.

   ■ Find the equations of the shapes (if any) where the surface intersects each coordinate plane.

   ■ Find the points (if any) where the surface intersects each coordinate axis.

   b. Translate the sphere with equation $x^2 + y^2 + z^2 = 16$ so that its center is at the point $P(4, -2, 3)$.

   ■ Sketch both spheres.

   ■ Write the equation for the translated sphere in expanded form.

   c. Formulate an equation for a sphere with center at (a, b, c) and radius r.

   d. Describe as completely as possible the surface with the equation
   $x^2 + y^2 + z^2 - 2x - 8z = 8$.

3. In Course 3, you saw that transforming rules of functions of a single variable transformed their graphs in predictable ways. Shown at the right is a graph of the paraboloid $z = x^2 + y^2$. Describe the graph of each of the following functions, and describe its relationship to the original paraboloid. If possible, check your predictions using three-dimensional graphing software or an appropriate calculator.

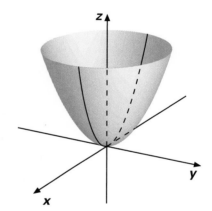

   a. $z = x^2 + y^2 + 5$

   b. $z = 3 - x^2 - y^2$

   c. $z = x^2 + (y - 2)^2$

   d. $z = 5x^2 + 5y^2$

4. The disposable drinking cup shown at the right can be thought of as a surface of revolution with a bottom. The cup has a 5-cm diameter bottom, 7-cm diameter top, and has height 9 cm.

   a. Beginning with an appropriate segment in three-dimensional space, describe how the surface can be generated.

b. The volume of the cup can be approximated by cutting it horizontally into sections each with a height of 0.5 cm, approximating each section with a cylinder and then summing the volumes of the cylinders. Approximate the volume in this manner.

c. Another way to generate the cup surface is to make the rotating segment part of a line through the origin. In this case, the volume of the cup is the difference of the volumes of two cones. Find the volume in this manner.

d. Compare the values of the volume found in Parts b and c. Describe how you could improve your approximation in Part b.

Reflecting

1. Suppose you are standing at the point $S(4, 5, -3)$. North is in the negative x-direction, east is in the positive y-direction, and up is in the positive z-direction.

   a. At what coordinates will you be standing after you move north 6 units, up 5 units, west 8 units, down 3 units, east 10 units, and south 4 units? Call this point A.

   b. After completing the trip in Part a, will you be looking up or down at the point $T(-2, 5, 1)$?

   c. After completing the trip in Part a, will you be closer to point S or point T?

2. In sketching a surface, why is it important to examine other cross sections, in addition to the traces?

3. Sketch graphs of $z = 0$, $z = 5$, and $z = -2$ on the same three-dimensional coordinate system. Describe how the graphs are related.

4. In this lesson, you found reasoning by analogy helpful in extending ideas of two-dimensional space to three-dimensional space. In a science course, you may have learned that Einstein's theory of special relativity involves a four-dimensional model (x, y, z, t). In that model, x, y, and z represent the usual coordinates of three-space and t represents the time coordinate. Think about how you could generalize the following ideas to four-dimensional space:

   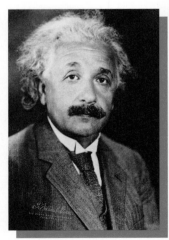

   Albert Einstein

   a. Distance between points $P(x_1, y_1, z_1, w_1)$ and $Q(x_2, y_2, z_2, w_2)$;

   b. Midpoint of $\overline{PQ}$ for the points in Part a;

   c. Analog of a sphere with center at $(0, 0, 0, 0)$ and radius r.

d. In two-space, a square can be formed by translating a segment of length s a distance of s units in a direction perpendicular to the segment, and then connecting corresponding vertices. In three-space, a cube can be found by translating a square of side length s by a distance of s units in a direction perpendicular to the plane of the square, and then connecting the corresponding vertices. How could you form a *hypercube*, the four-dimensional analog of a cube? How many vertices would the hypercube have? How many edges?

Extending

1. The coordinates of the midpoint of $\overline{AB}$ where $A(x_1, y_1, z_1)$ and $B(x_2, y_2, z_2)$ are $\left(\frac{x_1 + x_2}{2}, \frac{y_1 + y_2}{2}, \frac{z_1 + z_2}{2}\right)$. Suppose M is on $\overline{AB}$ and divides $\overline{AB}$ so that $AM = \frac{m}{n}AB$, where $0 < m < n$. Find the coordinates of this point of division. Check that when $\frac{m}{n} = \frac{1}{2}$, the coordinates are those of the midpoint of $\overline{AB}$.

2. A system of two linear equations in two variables represents two lines in a plane. The two lines may intersect in 0, 1, or infinitely many points. You have used several strategies to solve such systems, including the inverse-matrix method, linear-combinations method, substitution method, and graphing. In this task, you will consider the interpretation and solution of systems of linear equations in three variables.

a. What are the possible intersections for the graphs of two linear equations in three variables?

b. What are the possible intersections for the graphs of three linear equations in three variables?

c. How would you solve the following system of equations by extending the inverse-matrix method to three-space? Solve the system and check your solution.

$$(1) \quad x + y + z = 1$$
$$(2) \quad 2x - y + z = 0$$
$$(3) \quad x + 2y - z = 4$$

d. Solve the system in Part c by extending the linear-combination method for two-space. Begin by combining equations (1) and (3) and equations (2) and (3) to eliminate the z-variable. Then solve the resulting system of two variables. Compare your solution to that obtained in Part c.

e. Solve the system in Part c by first using the linear-combination method to eliminate the y-variable.

f. Solve the following system by using the linear-combination method and then check your result using the inverse-matrix method.

$$x + 2y + 2z = 5$$
$$x - 3y + 2z = -5$$
$$2x - y + z = -3$$

3. Find the equation of the plane through $P(2, 0, 0)$, $Q(0, 5, 0)$, and $R(0, 0, -8)$.

4. Many of the surfaces you have sketched in this unit have descriptive names. Most have three symbolic forms, each related to a particular axis, and involving combinations of quadratic terms, linear terms, and constants. Match each equation below with the appropriate surface. Then explain why your match makes sense.

a. $x^2 + y^2 + z^2 = r^2$

b. $\dfrac{x^2}{a^2} + \dfrac{y^2}{b^2} + \dfrac{z^2}{c^2} = 1$

c. $\dfrac{x^2}{a^2} + \dfrac{y^2}{a^2} - \dfrac{z^2}{b^2} = 0$

d. $\dfrac{x^2}{a^2} + \dfrac{y^2}{b^2} - \dfrac{z^2}{c^2} = 1$

e. $\dfrac{z^2}{c^2} - \dfrac{x^2}{a^2} - \dfrac{y^2}{b^2} = 1$

f. $\dfrac{x^2}{a^2} + \dfrac{y^2}{b^2} - z = 0$

g. $\dfrac{y^2}{b^2} - \dfrac{x^2}{a^2} - z = 0$

i. Ellipsoid

ii. Hyperboloid of one sheet

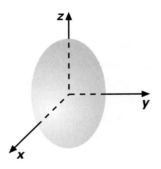

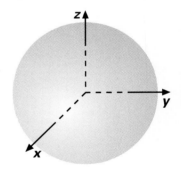

iii. Elliptic Paraboloid

iv. Sphere

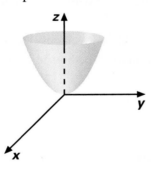

v. Hyperbolic Paraboloid

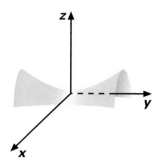

vi. Hyperboloid of two sheets

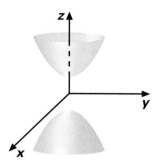

vii. Double Cone

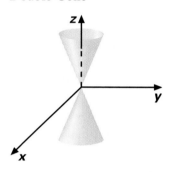

5. You have seen that the graph of an equation in three variables is not always easily visualized. Graphing technology is helpful in displaying such images, but even when it is used, you must manipulate the viewing window to obtain a clear understanding of the shape of the surface.

a. What window settings would you use to see the trace in the *yz*-plane? In the *xz*-plane? In the *xy*-plane?

b. A researcher involved in a study to determine the life of a new battery, measured in terms of charge-discharge cycles, found that battery life *Z* was a function of the charge rate *X* and the temperature where it was used, *Y*. The function was

$$Z = 262.58 - 55.83X + 75.50Y + 27.39X^2 - 10.61Y^2 + 11.50XY.$$

Use three-dimensional graphing software or an appropriate calculator to examine enough views of the surface to be able to describe the effect of charge rate and temperature on battery life.

1. If $\frac{-x}{14} = \frac{-7}{3}$, then $x =$

   (a) $-\frac{98}{3}$ (b) $\frac{85}{3}$ (c) $\frac{98}{3}$ (d) 98   (e) 33

2. Simplify: $\dfrac{3}{x + \frac{1}{9}}$

   (a) $\frac{3}{x+1}$ (b) $\frac{27}{x}$ (c) $\frac{30}{x+1}$ (d) $\frac{x+1}{30}$ (e) $\frac{27}{9x+1}$

3. Find the distance between $P(-2, 4)$ and $Q(6, -2)$.

   (a) 100   (b) $\sqrt{10}$ (c) 8   (d) 14   (e) 10

4. Let $f(x) = x^2 + 5$. Find $\frac{f(1 + h) - f(1)}{h}$.

   (a) $h + 2$ (b) $h^2 + 2h$ (c) $h^2 + 5$ (d) $2h + 6$ (e) $h - 2$

5. Solve $\sin^2 x + \sin x - 2 = 0$ in the interval $0 \le x < 2\pi$.

   (a) $x = \frac{\pi}{2}$ or $x = \frac{3\pi}{2}$ (b) $x = \frac{\pi}{2}$ (c) $x = \pi$

   (d) $x = \frac{\pi}{2}$ or $x = -2$ (e) $x = -2$ or $x = 1$

6. Solve for y: $ax = ay + \frac{1}{3}w$

   (a) $y = a\left(x - \frac{w}{3}\right)$ (b) $y = ax - a - \frac{1}{3}w$ (c) $y = x - \frac{w}{3}$

   (d) $y = x - \frac{w}{3a}$ (e) $y = \frac{w}{3a} - x$

7. The graph shows a polynomial function $y = f(x)$. Which graph below represents $y = f(x) + 2$?

$y = f(x)$

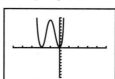

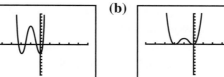 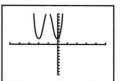

(a) **(b)** **(c)**

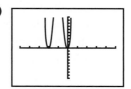

(d) **(e)**

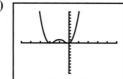

8. If b is 7 units and θ is 30°, find the height, h, of the triangle.

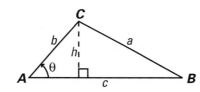

(a) 6.9 **(b)** 3.5 **(c)** 3 **(d)** 4 **(e)** 6.1

9. Write using logarithmic notation: $a^b = c$.

(a) $\log_b a = c$ **(b)** $\log_a b = c$ **(c)** $\log_a c = b$

(d) $\log_c a = b$ **(e)** $\log_b c = a$

10. Solve: $7 - 6\sqrt{x} = -11$.

(a) no real solutions **(b)** $x = 121$ **(c)** $x = -11$

(d) $x = 9$ **(e)** $x = -121$

Looking Back

In this unit, you developed skill in interpreting and representing surfaces in three dimensions. In the process, you investigated the usefulness of contour diagrams and cross sections as means to represent and analyze complex three-dimensional shapes. Analysis of cross sections of a double cone led to an alternate way of thinking about special curves you had previously studied—the circle, ellipse, parabola, and hyperbola.

The introduction of a three-dimensional coordinate system enabled you to explore and formalize analogs of important geometric ideas in two-space including distance, symmetry, and graphs of equations. Although three-dimensional graphing software and calculators with similar capabilities enable you to produce graphs of three-dimensional surfaces, being able to visualize the cross sections and intercepts was essential in interpreting the images. In this final lesson, you will review and consolidate your understanding of these key concepts and skills.

1. The Los Caballos Trail in the White Mountains in Arizona is a loop trail for hiking, mountain biking, and horseback riding. The trail is about 14 miles long. Use the information in the contour map at the top of the next page to help complete the following tasks.

   a. Estimate the altitudes of the Joe Tank and the Morgan Tank.

   b. Estimate the altitude of the trail head at the star.

   c. Forest Road 140 (FR 140) drops steadily between FR 300 and FR 136. How far does it drop?

   d. Which parts of the trail show the greatest changes in elevation?

   e. What do the closed curves in the upper left-hand portion of the map represent?

   f. Explain why the contour lines in this map and those in other contour diagrams in this unit never cross each other.

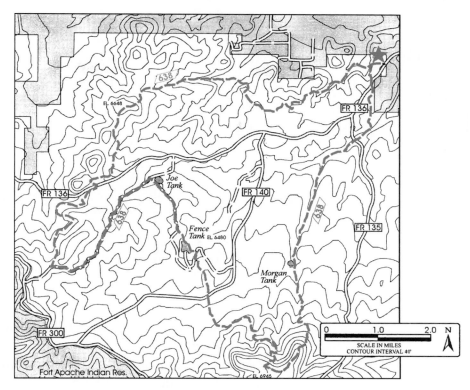

2. Prior to the Pathfinder mission and landing of the rover (shown below) on Mars in July 1997, research scientists worked with altitude data from earlier satellite space probes. The altitude data below is from a space probe of the region of Mars near Tiu Vallis at 8° N, 24° W (UTM −46,080 × 15,360). Data points were taken at 180-foot intervals.

4792	4292	3998	3998	3996	3996	3996	3998	3998
4938	4534	4128	3998	3998	3998	3998	3998	4248
4740	4518	4294	4070	3998	3998	3998	4248	4748
4666	4512	4358	4222	4154	4136	4236	4718	5038
4626	4508	4392	4336	4314	4352	4484	4808	5042
4618	4524	4472	4464	4484	4592	4750	5018	5122
4606	4538	4548	4562	4602	4728	4926	5086	5186
4610	4622	4648	4688	4770	4932	5082	5196	5292
4678	4710	4752	4822	4942	5086	5206	5316	5402
4752	4796	4858	4954	5086	5216	5332	5434	5512

a. Choose an appropriate contour interval for drawing a contour map of the region using these data. Draw the corresponding map.

b. Write a description of this region of Mars.

c. Sketch and describe a vertical cross section from left to right across the middle of the region.

3. Identify the conic section represented by each equation or graph below.

   ■ In the case of an equation, rewrite the equation in standard form and use properties revealed by the equation to sketch the graph.

   ■ In the case of a graph, write the corresponding equation.

   a. $9x^2 - 16y^2 = 144$

   b. $2x^2 + 2y^2 - 8x + 12y - 36 = 0$

   c.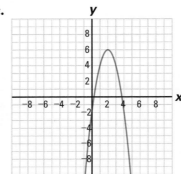

   d.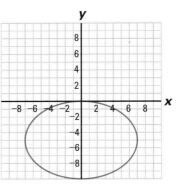

4. Complete the following tasks in the context of a three-dimensional coordinate system.

   a. Suppose you are standing above the xy-plane at the point $P(5, 8, 3)$ and looking at the point $Q(1, -3, 5)$. Are you looking up or down? How far is point Q from point P?

   b. Describe the set of points whose distance from the y-axis is 4. What equation describes this set of point(s)?

   c. Which of the points $A(3, 1, -2)$, $B(2, -3, 0)$, or $C(-10, 0, 0)$ is closest to the yz-plane? Which point(s) are on the xy-plane?

   d. Find the equation of a sphere of radius 5 with center at the origin. What is the equation of a congruent sphere with center at $(1, -3, 2)$?

5. Consider the plane with equation $6x + 4y + 3z = 24$.

   a. Sketch the plane.

   b. How would you describe the horizontal cross sections of this plane?

   c. Write the equation of a plane parallel to the given plane.

   d. Write the equation of a plane that intersects the given plane at its yz-trace.

6. A surface has the equation $x^2 + 4y^2 + 16z^2 = 64$.

   a. Find the equations of the traces.

   b. Find the intercepts.

   c. Describe all symmetries.

   d. What curves do planes perpendicular to the z-axis form with the surface?

   e. Sketch the surface.

7. Match each equation with one of the following surfaces.

  a. $4x^2 + 9y^2 = 36z$

  b. $36x^2 + 16y^2 - 9z^2 = 144$

  c. $12x + 2y - 9z = 36$

  d. $9(x - 3)^2 + 4(y - 6)^2 + 2(z - 4)^2 = 36$

  e. $6x + 4y = 24$

i.
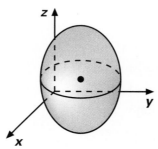

ii.
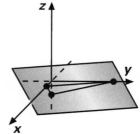

iii.
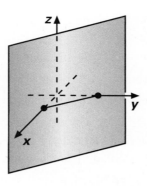

iv.
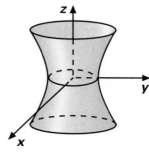

v.
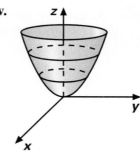

8. Which of the surfaces in Task 7 can be interpreted as surfaces of revolution about a coordinate axis? Explain your reasoning.

Checkpoint

In this unit, you added new methods for representing and analyzing surfaces to your toolkit. This included extending coordinate methods of a plane to three-dimensional space.

ⓐ Describe how to use altitude data to draw a contour map of the surface of a region.

ⓑ Conic sections are special intersections of a plane with the surface of a double cone. For each conic section:

- Describe geometrically how it is formed.
- Describe how it can be defined in terms of a specified distance condition in a plane.
- Give its standard form algebraic representation.
- Describe an application of the curve.

ⓒ Compare the manner in which the following concepts are modeled in two-space and in three-space:

- point
- distance
- midpoint
- plane
- symmetry
- quadratic equations

ⓓ What strategies would you use to sketch a graph or check the reasonableness of a technology-produced graph of a three-dimensional surface represented by an equation in three variables? In two variables?

ⓔ Describe and illustrate cylindrical surfaces and surfaces of revolution.

Be prepared to share your descriptions and thinking with the entire class.

▶On Your Own

Write, in outline form, a summary of the important mathematical concepts and methods developed in this unit. Organize your summary so that it can be used as a quick reference in future work.

Informatics

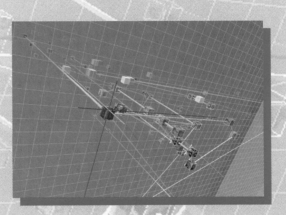

Lesson 1

Access—The Mathematics of Internet Search Engines

You are living in the midst of an ongoing revolution in information processing and telecommunications. Telephones, televisions, and computers are merging. You may listen to digital music, watch digital movies, and use an ATM machine to do digital banking. More and more transactions are taking place on the Internet. How is all this digital information transmitted and received in an accurate and secure manner? To help answer this question, you will study the mathematics of information processing, called **informatics**. You will consider three fundamental issues of informatics—access, security, and accuracy. Each issue will be the focus of a lesson in this unit.

One of the most important applications of informatics is the Internet.

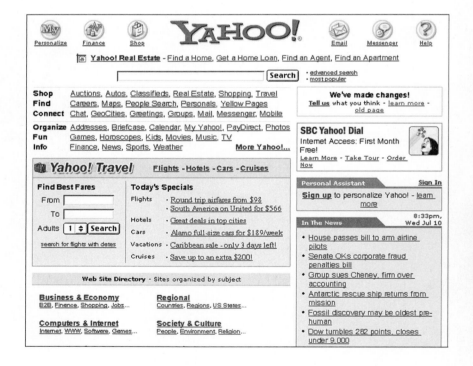

INVESTIGATION 1 ▶ Set Theory and the Internet

Businesses, libraries, schools, and governments collect masses of data, words, pictures, and sounds, and the Internet makes much of that information available around the world at any hour of the day or night. An important branch of mathematics that is used to help efficiently store and retrieve information is *set theory.* Roughly speaking, a *set* is a collection of objects. There are many English words used to describe collections of things. We talk about families of people, flocks of birds, forests of trees, bands of musicians, and fleets of ships. To facilitate clear communication, mathematicians generally agree to use the single word **set** to talk about any well-defined collection of objects. While a precise treatment of sets requires setting up a system of axioms (assumptions), this informal definition of a set is sufficient for now. The individual members of a set are called its **elements**. It doesn't matter in which order the elements of a set are listed.

1. Sets are represented by listing or describing the elements of a set within braces (curly brackets). The elements can be any "objects." For example, all of the following are sets:

   $N = \{2, 3, 5, 7, 11, 13\}$ $\qquad\qquad$ $P = \{5, \text{exponent}, \text{cube}, \text{dog}\}$
   $S = \{\text{triangle}, \text{square}, \text{point}, \text{prism}, \text{icosahedron}\}$
   $W = \{\text{illuminations.nctm.org}, \text{www.wmich.edu/cpmp}, \text{mathforum.org}\}$

   If an element x is in a set A, write $x \in A$, which is read as "x is an element of the set A" or "x is in A." To indicate that "x is *not* an element of the set B," the notation $x \notin B$ is used. Two sets are said to be **equal sets** if and only if they have the same elements.

   a. How many elements are in each of sets N, P, S, and W above?

b. Is {2, 3, 4} = {3, 4, 2}? Explain.

c. Give an example of a set A that has the following properties: $5 \in A$, $10 \notin A$, and A has five elements.

2. The elements in a set often share a common property. In such cases, you can define a set in terms of that property. There is specific notation, called *set-builder notation*, that is used to do this. For example,

$A = \{x \,|\, x$ is an odd integer greater than 2 and less than 10$\}$ is read:

"A equals the set containing *all* elements x such that x is an odd integer greater than 2 and less than 10." Thus, $A = \{3, 5, 7, 9\}$.

a. List the elements in each of the sets below.

- $S = \{x \,|\, x$ is an integer and $-2 \leq x \leq 5\}$
- $P = \{x \,|\, x = 2^n$, where $n = 0, 1, 2,$ or $3\}$
- $M = \{x \,|\, x$ is the name of a month that has exactly 30 days$\}$

b. Write each of the following sets using set-builder notation.

- $S = \{1, 4, 9, 16, 25\}$
- $E = \{2, 4, 6, 8, 10, \dots\}$
- $P = \{$triangle, quadrilateral, pentagon, hexagon$\}$

Set theory can be used to narrow an Internet search to a manageable number of sites. For example, a Web search in July, 2002 on the topic "Music" reported about 71,700,000 possible sites with relevant information.

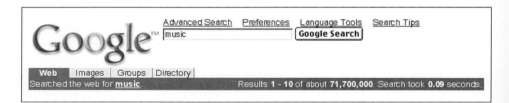

In making a Web search, you are defining a set—the collection of those Web sites that have some common feature in which you are interested. There are three *logical operators*, AND, OR, and AND NOT, that help in focusing your search on the sites that might be most useful for a particular project.

3. What kinds of information would you expect to find from searches that look for Web sites meeting these combinations of conditions:

a. music AND history

b. music OR history

c. music AND NOT videos

d. rap OR rock AND music

e. rap AND classical AND music

4. Episodes I, II, and III of the *Star Wars* saga have been perhaps the most eagerly anticipated movies in recent years. There are many Web sites that have information about the *Star Wars* saga and its characters.

Star Wars fans await the premier of Episode II.

   a. What search conditions could you use to find information about the actors who played Anakin in the *Star Wars* saga?

   b. How could you design a search to find information about box office receipts from the various *Star Wars* movies.

   c. How could you design a search to get information comparing *Star Wars* and *Star Trek* movies?

When you use the words AND, OR, and AND NOT to define searches on the World Wide Web, what you are really doing is defining new sets of information that are combinations of existing sets.

Information about *Star Wars* or any other topic is found on the Web by searching through lists of Web pages. There are two ways that these lists are created—by people or by automated search engines. For a Web directory like Yahoo!®, or About.com, people create at least some of the listings. You can submit a short description of a Web site that you would like listed, but not every Web site submitted will be listed. That is a decision made by the editors. The editors also include descriptions for selected sites that they review. When a user searches a Web directory, the search only looks for matches in the descriptions listed.

Another way that lists of Web pages are created is by automated search engines. An automated search engine, like those used by AltaVista® or Google™ has three basic components. A *spider*, or *crawler*, visits many Web pages, reads them, and returns on a regular basis to look for changes. Information gathered from the spider's visits goes into the search engine's *index*, which is like a giant book that contains a summary of every page that the spider finds. The third part of a search engine is the *searching software* that sifts through the millions of pages recorded in the index to find matches for your search.

The searching software almost instantaneously finds Web pages relevant to what you are looking for, and ranks them in order of relevance. Of course, search engines don't always get it right, but their speed and accuracy is pretty amazing.

There are sophisticated and proprietary methods that each individual search engine software uses, but they are all based on the same fundamental mathematics. Consider the following information, found on an AltaVista Help page in July, 2002. Note the first three logical or *Boolean operators* at the top of the following Web page.

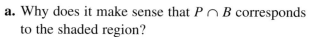

altavista
THE SEARCH COMPANY

Home > Help > Search > **Special search terms**

You can use these terms for both <u>basic</u> and <u>advanced</u> Web searches. For advanced searches, type these into the free-form Boolean box.

AND	Finds documents containing all of the specified words or phrases. **Peanut AND butter** finds documents with both the word peanut and the word butter.
OR	Finds documents containing at least one of the specified words or phrases. **Peanut OR butter** finds documents containing either peanut or butter. The found documents could contain both items, but not necessarily.
AND NOT	Excludes documents containing the specified word or phrase. **Peanut AND NOT butter** finds documents with peanut but not containing butter. NOT must be used with another operator, like AND. AltaVista does not accept **'peanut NOT butter'**, instead, specify **peanut AND NOT butter**.
NEAR	Finds documents containing both specified words or phrases within 10 words of each other. **Peanut NEAR butter** would find documents with peanut butter, but probably not any other kind of butter.
*	The asterisk is a wildcard; any letters can take the place of the asterisk. **Bass*** would find documents with bass, basset and bassinet. You must type at least three letters before the *. You can also place the * in the middle of a word. This is useful when you're unsure about spelling. **Colo*r** would find documents that contain color and colour.
( )	Use parentheses to group complex Boolean phrases. For example, **(peanut AND butter) AND (jelly OR jam)** finds documents with the words 'peanut butter and jelly' or 'peanut butter and jam' or both.

5. According to the special search terms table above, if you search for "peanut AND butter" you will find documents that contain both the word "peanut" and the word "butter." This corresponds to the set operation called *intersection*. The **intersection** of sets A and B is the set containing all elements that are in *both A and B*. This is denoted $A \cap B$ and is read "A intersect B." Thus,

$$A \cap B = \{x \mid x \in A \text{ and } x \in B\}.$$

Using sets and set intersection you can represent the search for "peanut AND butter" as follows:

$P = \{x \mid x \text{ is a document containing the word "peanut"}\}$

$B = \{x \mid x \text{ is a document containing the word "butter"}\}$

$P \cap B = \{x \mid x \text{ is a document containing both the words "peanut" and "butter"}\}$

Venn diagrams (named for the nineteenth-century British mathematician John Venn) are a useful pictorial way to represent sets and intersections. The shaded region in the Venn diagram at the right represents $P \cap B$.

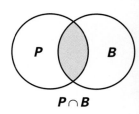

$P \cap B$

a. Why does it make sense that $P \cap B$ corresponds to the shaded region?

b. Suppose that you wanted to do an Internet search for colleges in Minnesota. Represent this search using set notation and using a Venn diagram.

c. Suppose that $C = \{1, 3, 4, 6, 8\}$ and $D = \{2, 3, 4, 8, 10, 12\}$. Find $C \cap D$. Draw a Venn diagram that illustrates this situation.

6. According to the special search terms table on page 582, if you search for "peanut OR butter" you will find documents containing either "peanut" or "butter." The found documents could contain both words, but that is not required. This corresponds to the set operation called *union*. The **union** of sets A and B is the set containing all elements that are in *either A or B* (or both). This is denoted $A \cup B$ and is read "A union B." Thus,

$$A \cup B = \{x | x \in A \text{ or } x \in B\}.$$

Venn diagrams can also be used to represent the union of sets. The shaded region in the Venn diagram at the right represents $A \cup B$.

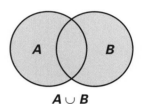

$A \cup B$

a. Why does this representation make sense?

b. Represent the search for "peanut OR butter" using set notation and using a Venn diagram.

c. Suppose that $C = \{1, 3, 4, 6, 8\}$ and $D = \{2, 3, 4, 8, 10, 12\}$. Find $C \cup D$. Draw a Venn diagram that illustrates this situation.

d. Suppose that you are shopping for a used sports car or convertible. You have entered a used-car-shopping Web site that has a built-in search engine. Describe the search you would do to find the types of cars in which you are interested. Describe the search in three ways: using a logical (Boolean) expression, using the union of sets, and using a Venn diagram.

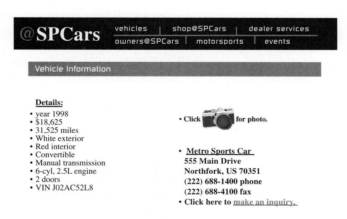

e. Suppose that a family is considering purchasing an SUV (sport-utility vehicle) or a minivan. They decide to research both types of vehicles. Draw a Venn diagram that illustrates a search in this situation.

7. You can form unions and intersections of any number of sets. Consider the search example below.

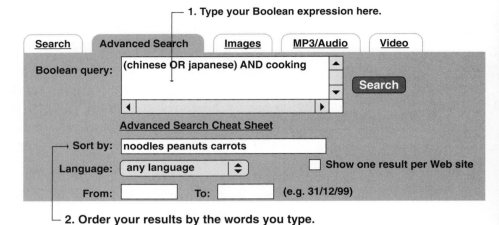

1. Type (chinese OR japanese) AND cooking in the Boolean query box. This Boolean expression will find all the documents containing Chinese cooking or Japanese cooking.

2. Type noodles peanuts carrots in the Sort by box. This step orders the cooking documents so that noodles peanuts and carrots appear at the top of your results.

3. Press the Search button.

a. There are three steps listed in this search example. Represent the situation in Step 1 using union and intersection. Draw a Venn diagram that illustrates this situation.

b. After completing the search shown, suppose you now want to find documents that include all three of these ingredients: noodles, peanuts, and carrots. Represent such a search using a Boolean expression. Also represent this search with sets and illustrate the situation with a Venn diagram.

8. According to the special search terms table on page 582, if you search for "peanut AND NOT butter" you will find documents that contain the word "peanut" but do not contain the word "butter." This corresponds to the set operation called *set difference*. The **set difference** of sets A and B is the set containing all elements that are in A and are not in B. This is denoted $A - B$ and is read "the difference of A and B."

a. Suppose that $X = \{1, 3, 4, 6, 8\}$ and $Y = \{2, 3, 4, 8, 10, 12\}$. Find $X - Y$.

b. Write the definition of $A - B$ using set-builder notation by completing the statement below.

$$A - B = \{x \mid \qquad \}$$

c. Draw a Venn diagram that illustrates $A - B$.

d. Use sets and the set difference operation to represent the search for "peanut AND NOT butter." Draw a Venn diagram that illustrates this situation.

e. Use sets and the set difference operation to represent the search for "butter AND NOT peanut." Draw a Venn diagram that illustrates this situation.

9. Joaquin's favorite types of music are Rock and Blues. He is planning an Internet search to find information about these two types of music.

a. Use three sets to represent a search that will find information about these types of music. Draw a Venn diagram that illustrates this situation.

b. Joaquin is particularly interested in "crossover" artists who play both Rock and Blues. Draw a Venn diagram that represents this situation.

c. Joaquin's friend Tionne is a Blues purist. She prefers Blues artists who do not incorporate Rock. She does a search for such artists. Represent this search using sets and the appropriate set operation. Draw a Venn diagram that represents this situation.

d. If you were actually doing an Internet search for these types of music, what search results might arise unrelated to your request? How could you modify your Internet search to reduce unwanted results?

Checkpoint

In this investigation, you learned about union and intersection of sets and set difference, and how these ideas are helpful in Internet searches.

a What are the two common notational methods for defining a set?

b Logical operators like AND, AND NOT, and OR are used in many Internet search engines. Describe the relationship between each of these operators and set operations. For each of the three operators, describe an Internet search that would involve the operator.

c Shade regions on copies of this Venn diagram to represent each of the following sets:

- $A \cap B \cap C$
- $A \cup B \cup C$
- $A \cup (B - C)$

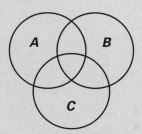

Be prepared to share your descriptions and Venn diagrams with the entire class.

On Your Own

The operations union, intersection, and difference combine two sets to give another set.

a. Describe Internet searches that have not already been discussed that would involve the intersection of three sets; the difference of two sets.

b. Consider the following sets.

$$A = \{2, 4, 6, 8, 3, 17, \pi\} \qquad B = \{1, 3, 5, 7, 2, 8\}$$

■ Find $A \cap B$, $A \cup B$, $A - B$, and $B - A$. Draw Venn diagrams that illustrate these four set operations.

■ Let $E = \{x \mid x \text{ is an even number}\}$. Find $A \cap E$ and $A - E$. How are $A \cap E$ and $A - E$ related?

There are some other relations involving sets that are not reflected in the special search terms table, but that are nevertheless important in many situations. You will investigate some of these ideas about sets in the following activities.

10. Another common set operation is *set complement*. The **complement** of a set is everything in the "universe" that is not in the set. The "universe" is defined by the **universal set**, which is the set of all objects that are under consideration. That is, the universal set defines the overall context. The complement of a set A is denoted as A' and is read "A complement" or "the complement of A."

a. Let the universal set be $U = \{5, 10, 15, 20, 25, 30, 35\}$. Suppose $A = \{10, 20, 30\}$ and $B = \{5, 10, 15, 20\}$. Find A' and $(A \cap B)'$.

b. Consider the universal set $Z = \{x \mid x \text{ is an integer}\}$. Find the complement of the set $E = \{x \mid x \text{ is an even integer}\}$.

c. You can illustrate the universal set in a Venn diagram by drawing a large rectangle around the circles that represent the sets on which you are operating. Draw Venn diagrams that illustrate the situations in Parts a and b.

11. Use set operations to describe the set indicated by each Venn diagram.

a. 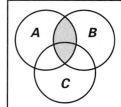   **b.** 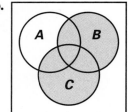   **c.**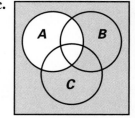

12. A set A is a **subset** of set B, denoted $A \subseteq B$, if and only if every element of A is also an element of B. Consider the set $N = \{2, 4, 5, 6, 8, 11\}$.

a. Is $\{5, 2\}$ a subset of N? Why or why not?

b. Is $\{4, 6, 8, 10\} \subseteq N$? Why or why not?

c. Write two other subsets of *N*.

d. Is {2, 4, 5, 6, 8, 11} a subset of *N*? Justify your answer by using the definition of "subset."

e. In a search for used Chevrolet trucks on the Internet, think about three possible searching sets:

$T = \{x \mid x$ is a document containing the word "truck"$\}$

$U = \{x \mid x$ is a document containing the phrase "used truck"$\}$

$C = \{x \mid x$ is a document containing the word "Chevrolet"$\}$

Which of these sets is a subset of another set? Draw a Venn diagram that shows the relationships among these three sets.

13. A subset of a set that does not equal the set is called a **proper subset**. To indicate that a set *A* is a proper subset of a set *B*, write $A \subset B$.

a. Identify those subset relationships in Activity 12 that are proper subset relationships.

b. Now consider $A \cup B$ and $A \cap B$, where *A* and *B* are any two sets. Is either $A \cup B$ or $A \cap B$ *always* a subset of the other? As you answer this question, consider the following questions and draw Venn diagrams to illustrate your claims.

   ■ Is one a *proper* subset of the other?

   ■ What happens if $A = B$?

   ■ What happens if *A* is a subset of *B*?

c. Next consider the sets $A - B$ and *A*, where *A* and *B* are any two sets. Is $A - B$ or *A* *always* a subset of the other? As in Part b, consider subsets and proper subsets and draw Venn diagrams to illustrate your claims.

14. The set with no elements is called the **empty set** or **null set**, denoted by $\varnothing$. Thus, $\varnothing = \{\ \}$.

a. Describe an Internet search whose result would probably be the empty set.

b. In Unit 4, "Counting Models," the empty set was considered a subset of every set. This may seem a bit strange, but it follows from the definition of a subset given in Activity 12. Prove that the empty set is a subset of every set. (*Hint:* One way to prove this is to think about how you would show that a set *A* is *not* a subset of a set *B*.)

15. Two sets that do not share any elements are said to be **disjoint**. That is, sets *A* and *B* are *disjoint* if and only if $A \cap B = \varnothing$. State and explain whether the sets in each of the following pairs are disjoint. If the sets are not disjoint, draw a Venn diagram and describe the elements that are in the intersection.

a. {1, 2, 4, 7, 9, 13, 55, 607} and {−3, −5, 0, 7}

b. $\{x \mid x$ is an even number$\}$ and $\{x \mid x$ is an odd number$\}$

c. $\{x \mid x$ is a Web page containing the word "mathematics"$\}$ and $\{x \mid x$ is a Web page containing the word "poetry"$\}$

d. $A - B$ and $B - A$

Checkpoint

In this lesson, you learned about sets, set operations, and relationships between sets.

ⓐ Suppose $X = \{1, 2, 3, 4, 5\}$. Which of the following statements are true? Explain.

- $3 \in X$
- $\{3\} \in X$
- $3 \subseteq X$
- $\{3\} \subseteq X$

ⓑ If P, Q, R, and S are all subsets of a universal set U, what do each of the following expressions tell about the relationships among the sets?

- $P \cup R = P$
- $Q \cap R = R$
- $S \subseteq P'$
- $S - R = S$

ⓒ Using the ideas of *subset* and *disjoint*, describe the relationships among $A \cap B$, $A \cup B$, and $A - B$. Draw Venn diagrams to illustrate the relationships.

ⓓ The main theme of informatics that you considered in this lesson is "access." Explain why this issue is important in informatics. Explain what the topics you learned about in this lesson have to do with "access."

Be prepared to share your thinking and Venn diagrams with the entire class.

▶ On Your Own

Check your understanding of set operations and relationships by completing the following tasks.

a. Use Venn diagrams similar to the one at the right to shade regions representing each of the following sets.

- $A \cup B'$
- $A' \cap B$
- $A' \cap B'$

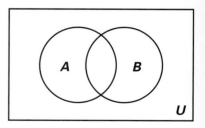

b. Let $A = \{3, 5, 10\}$.

- List all subsets of A.
- Which of the subsets that you listed are proper subsets?
- From the subsets that you listed, find two subsets that are disjoint.

MORE

Modeling • Organizing • Reflecting • Extending

Modeling

1. The inventory system of a used car superstore assigns each vehicle a code number. Information about each car is attached to that code number in the store's database, so a customer or salesperson can search the database to find cars with particular combinations of features. Suppose that a customer comes to the

store with preferences for three features in a car: Green color G; Low mileage L; and Two doors T. What sets of cars will be identified by searches for each of the following:

   a. $G \cap (L \cap T)$ **b.** $G \cap (L \cup T)$

   c. $L' \cup T'$ **d.** $(L \cup T)'$

2. Consider some Internet searches using logical operators AND, AND NOT, and OR.

   a. Use logical operators to design a search that will find information about Irish dancing. Represent the search using sets and set operations. Illustrate the situation with a Venn diagram.

   b. Use logical operators to design a search that will find information about Irish or English dancing. Represent the search using sets and set operations. Illustrate the situation with a Venn diagram.

   c. Describe an Internet search situation that involves the logical operator AND NOT.

3. Suppose that in a probability experiment you flip a fair coin and then roll a fair die, recording the combination of heads or tails on the coin and numbers showing on top of the die.

   a. Draw a tree diagram showing all possible outcomes of this two-stage experiment.

   b. List the possible outcomes of the experiment as a set S with elements labeled with letter-number pairs like h1 or t6.

   c. List the elements in the sets defined by these conditions:

   $H = \{x \in S \mid x$ shows an outcome of heads$\}$

   $E = \{x \in S \mid x$ shows an even number die toss$\}$

   d. Which of the following statements are correct?

     i. $t5 \in E$

    ii. $(H \cap E) \subseteq E$

   iii. $\{h3\} \subseteq E'$

   e. List the elements in the following sets:

      i. $H \cup E$

     ii. $H \cap E$

    iii. E'

    iv. $(H \cap E)'$

      v. $H - E$

    vi. $H \cup \varnothing$

   vii. $E \cap S$

4. A survey of political opinions asked 100 people whether they favored (1) public funding of private schools; (2) government control of health care costs; (3) increase in nonmilitary aid to developing countries around the world. Results of the survey showed:

   - Only 5 people favored all three actions.

   - 10 people opposed all three actions.

   - 15 favored only the increase in nonmilitary aid to developing countries.

   - 13 favored both public funding for private schools and health care cost controls.

   - 20 favored health care cost control and aid to developing countries.

   - A total of 48 favored health care cost control.

   - 25 favored only the public funding for private school action.

Use a copy of the Venn diagram at the right to model this situation and to help answer the following questions.

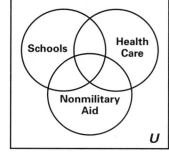

a. How many people favored at least one of the proposed actions?

b. How many people favored at least two of the proposed actions?

c. How many people favored either the health care cost control or aid to developing countries actions but not the public funding for private schools?

d. Use set notation to describe the set of people in each of Parts a–c, using S for the school issue, H for the health issue, and A for the aid issue.

Organizing

1. Consider the following sets:

$$A = \{2, 4, 6, 8, 10\}$$

$$B = \{1, 2, 3, 4, 5, 6\}$$

a. Find $A \cap B$, $A \cup B$, $B - A$, and $A - B$. Draw Venn diagrams that illustrate these four set operations.

b. List two subsets of A.

c. Let $C = \{x \mid x$ is an integer less than 10$\}$. Compare the sets A, B, and C. Which is a subset of which?

d. Write set A using set-builder notation.

2. Suppose that Q is the set of all quadrilaterals in a plane, S is the set of all squares, R is the set of all rectangles, P is the set of all parallelograms, and H is the set of all rhombuses. Which of the following statements correctly describes a relationship among these sets?

a. $S \subseteq R$

b. $R \cup H = P$

c. $R \nsubseteq P$ (The symbol $\nsubseteq$ means "is not a subset of.")

d. $R' = P$

e. $S \nsubseteq H$

f. $R \cap H = S$

g. $Q - P' = P$

h. $S \subseteq P$

3. Set union and intersection are *binary operations*, like addition and multiplication, in that they combine two sets to give another set, so it is natural to investigate whether those operations have properties like those of number systems. Consider the following three sets of numbers:

$$A = \{x \mid x \text{ is a whole number factor of } 42\}$$
$$B = \{x \mid x \text{ is a whole number factor of } 70\}$$
$$C = \{x \mid x \text{ is a whole number factor of } 105\}$$

List the elements of each set and then use the sets to test the following conjectures of algebraic properties for the binary operations $\cup$ and $\cap$.

For each statement that seems true, use a Venn diagram to provide a "proof without words" that the property will hold in general. For each statement that is false, provide a counterexample.

a. $(A \cup B) \cup C = A \cup (B \cup C)$

b. $(A \cap B) \cap C = A \cap (B \cap C)$

c. $(A \cup B) \cap C = A \cup (B \cap C)$

d. $A \cap (B \cup C) = (A \cap B) \cup (A \cap C)$

e. $A \cup (B \cap C) = (A \cup B) \cap (A \cup C)$

f. $(A \cap B)' = A' \cap B'$

4. Recall that a point in the coordinate plane is represented as an *ordered pair* (x, y) of real numbers. You can think about ordered pairs as elements of a set called the *Cartesian product*. The **Cartesian product** of two sets A and B is the set of all ordered pairs that have first coordinate from A and second coordinate from B. The Cartesian product of A and B is denoted by $A \times B$.

$$A \times B = \{(x, y) \mid x \in A \text{ and } y \in B\}$$

a. Let $A = \{2, 3, 4\}$ and $B = \{50, 60\}$.

■ Is $(2, 3) \in A \times B$? What about $(2, 50)$? What about $(60, 3)$?

■ The Cartesian product of A and B is partially written below. Complete the Cartesian product.

$$A \times B = \{(2, 50), (2, 60), (3, 50), \dots\}$$

b. Let $S = \{y, z\}$ and $T = \{1, 2, 3, 4\}$. Find $S \times T$.

c. How many elements are in $A \times B$ if A has n elements and B has m elements?

5. The following data show numbers of male and female students who were honors students in a school graduating class and numbers who were not honors students.

	Honors	Not Honors
Male	20	85
Female	25	70

Suppose that a student is chosen at random to be interviewed on a local television station about his or her experience in local schools. If M stands for the set of male students, H stands for the set of honors students, and x represents the chosen student, calculate each of the following probabilities:

a. $P(x \in M)$

b. $P(x \in M \cap H)$

c. $P(x \in M \cap H')$

d. $P(x \in M' \cup H)$

e. $P(x \in M | x \in H)$

f. $\dfrac{P(x \in M \cap H)}{P(x \in H)}$

Reflecting

1. The letters W, Z, Q, R, and C are commonly used as labels for the sets of whole numbers, integers, rational numbers, real numbers, and complex numbers, respectively.

   a. Draw a Venn diagram showing the relationships among these sets. The set of complex numbers is the universal set.

   b. Which of the following are correct statements about those number sets? If a statement is not correct, explain why not.

      i. $W = \{0, 1, 2, 3, 4, \ldots\}$

     ii. $W \in Z$

    iii. $Q = \{x \in R \,|\, x = \frac{a}{b},\, a \in Z,\, \text{and } b \in Z,\, b \neq 0\}$

     iv. $\sqrt{2} \in Q$

      v. $Q \subseteq R$

     vi. $\sqrt{-36} \in R'$

    vii. $\pi \in (R - Q)$

   viii. $W \cup Z = Q$

2. What familiar algebraic problems are indicated by these examples of set notation?

   a. Find $S = \{x \in R \,|\, 3x + 5 = 12\}$, where R is the set of real numbers.

   b. Find $S \cap T$, where $S = \{(x, y) \,|\, 3x + 2y = 8\}$ and $T = \{(x, y) \,|\, x - y = 4\}$.

   c. Find $S = \{(x, y) \,|\, x \geq 0,\, y \geq 0,\, x + y \leq 2,\, 3x + 2y \leq 4,\, \text{and } x + 2y \text{ is maximized}\}$.

3. Because of the importance of Internet search engines, a national Search Engine Meeting is held each year. The Seventh Search Engine Meeting was held in San Francisco on April 15–16, 2002. Go to http://infonortics.com/searchengines/ to find information about this meeting (or a more recent Search Engine Meeting). Choose one topic from the meeting that is most interesting to you and prepare a brief report on that topic.

4. Examine the following statement by Susan Landau, Associate Editor of the *Notices of the American Mathematical Society*, which is taken from an article entitled, "Internet Time."

> The information revolution will have as profound an effect on the world as the industrial revolution did. The organization and access of massive amounts of information is a deep and fundamentally mathematical problem.
>
> Source: *Notices of the American Mathematical Society*, Volume 47, Number 3, March 2000, p. 325.

Do you agree with Dr. Landau? Why or why not?

Extending

1. The **power set** of a set is the set of all subsets of that set. The power set of a set A is denoted by $P(A)$.

   a. Consider the set $A = \{a, b, c\}$. The power set of A is partially written below. Complete the power set.

   $$P(A) = \{\varnothing, \{a\}, \{b\}, \{c\}, \{a, b\}, \ldots\}$$

   b. Consider the set $A = \{a, b\}$. Find $P(A)$.

   c. Suppose A has four elements. How many elements are in $P(A)$? If A has n elements, how many elements are in $P(A)$?

   d. Complete the following argument to prove your conjecture in Part c. Be sure that you understand the reason supporting each step.

   Suppose $S = \{a_1, a_2, \ldots, a_n\}$:

   (1) For any subset, either a_1 is in the subset or not; this gives two choices.

   (2) For any subset and any choice of whether to include a_1 or not, there are two choices for a_2: put it in the subset or not.

   (3) In the same way for any element a_k and any choice of which of the preceding elements to include and which to exclude in a given subset, there are two choices of what to do with a_k: put it in the subset or not.

   (4) Therefore, for n elements, there are _____ possible choices for the individual elements to be included or not, so there are _____ elements in $P(S)$.

2. The most interesting questions about Cartesian product sets (Organizing Task 4) are those in which there is an important relationship between entries in some of the ordered pairs. In those situations, you want to know "Is x related to y in a particular way?" Formally, a **relation** between two sets A and B is a set of ordered pairs, where the first coordinate of the ordered pair is from A and the second coordinate is from B. Sometimes a relation is between two sets that are equal. In this case, the relation is called a *relation on A* (or on *B*).

   a. Describe the sets of points in a coordinate system that are designated by the following relations between sets of real numbers.

   ■ $\{(x, y) \mid y = 0.5x + 1\}$

   ■ $\{(x, y) \mid y < x\}$

   ■ $\{(x, y) \mid y > x^2\}$

   ■ $\{(x, y) \mid x^2 + y^2 = 7\}$

   b. Consider the "less than" relation on the set of all integers. That is, x is related to y if and only if x and y are integers and $x < y$. Since $3 < 4$, 3 is related to 4, and thus $(3, 4)$ is an element in the relation.

   ■ Is $(2, 9)$ in the "less than" relation? Is $(9, 2)$ in this relation? Is $(-4, -3)$?

   ■ Call this relation L. Thus, $L = \{(x, y) \mid x$ and y are integers and $x < y\}$. List three more elements of L. How many elements are in L?

   c. Let $A = \{a, b, c, d\}$. Let $T = \{t, u, v, w, z\}$. Refer to the formal definition of relation given above to answer the following questions.

   ■ Is $R = \{(a, t), (b, w), (c, z), (d, v)\}$ a relation between A and T?

   ■ Is $K = \{(a, t), (a, u)\}$ a relation between A and T?

   ■ Is $M = \{(a, t), (z, d)\}$ a relation between A and T?

   ■ Construct a relation between A and T that has 5 elements.

   d. How many possible relations are there between a set with 3 elements and a set with 2 elements? How about between a set with n elements and a set with m elements?

3. A *function* is a special type of relation. Informally, a function from set A to set B must "map" each element in A to exactly one element in B. For example, the function f defined by $f(x) = x^2$ maps each real number to one and only one nonnegative number.

   A function can be defined formally in terms of ordered pairs. A **function** F from set A to set B is a relation between A and B such that for every element x in A, there is exactly one ordered pair in F that has x as the first coordinate. (F maps each element in A to exactly one element in B.)

   There cannot be two ordered pairs in a function F that have the same first coordinates and different second coordinates. (A function cannot map the same element to two different images.)

Use this formal definition of function to answer the following questions.

a. Consider the function f defined by $y = x^2$. Formally, f is a set of ordered pairs. That is, $f = \{(x, y) \mid x$ and y are real numbers and $y = x^2\}$.

  ■ Is $(3, 9)$ in f? Is $(3, 6)$ in f?

  ■ Some of the elements of f are given below. Add three more elements to the set.

$$f = \{(3, 9), (-2, 4), (9, 81), \ldots\}$$

  ■ What is the domain of f? The range of f?

b. Consider the relations in Part a of Extending Task 2. Which of these relations is a function? Explain.

4. Two finite sets have the same number of elements if counting the sets produces the same result. Counting a set involves tagging each element with a positive integer starting with 1, 2, 3, and so on until all have been tagged.

a. Explain why the description of counting given above implies that any two sets with the same number of elements can be put in *one-to-one correspondence*. For example, $B = \{a, b, c, d, e, f\}$ and $E = \{z, y, x, w, v, u\}$ have the same number of elements and they can be matched in a one-to-one fashion (for instance, $a \leftrightarrow z$, $b \leftrightarrow y$, and so on).

b. For infinite sets, conventional counting is not possible. However, sometimes it is possible to show that two infinite sets have the same number of elements by showing that there is a one-to-one correspondence between them. Show in this way that the following pairs of infinite sets have the same number of elements:

  i. The positive integers and the negative integers

  ii. The positive integers and the even integers

  iii. The points on $\overline{MM'}$ and the points on $\overline{BC}$, where M and M' are the midpoints of $\overline{AB}$ and $\overline{AC}$, respectively

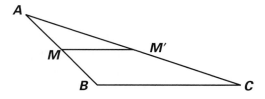

5. In this lesson, you investigated Internet searches using Boolean expressions. The term "Boolean" comes from the notion of a *Boolean algebra*, named in honor of its inventor George Boole (1815–1864).

George Boole

A Boolean algebra is a particular type of mathematical structure. Set theory provides an example of a Boolean algebra. First of all, a Boolean algebra must have some elements. In the case of set theory, the subsets of a set S are the elements of the Boolean algebra. There are two "special" elements, namely, the universal set S and the empty set. Secondly, there are three operations in a Boolean algebra. In the case of set theory, there are three relevant operations on the subsets—union, intersection, and complementation. Thirdly, the elements and operations must satisfy certain properties; namely, commutative, associative, distributive, identity, and complementation properties. Find a college discrete mathematics text that lists all the properties of a Boolean algebra. Prepare a report explaining why the subsets of a set S, along with the operations of union, intersection, and complementation, form a Boolean algebra.

1. What fraction of the circle is unshaded?

   (a) $\frac{2}{7}$ (b) $\frac{5}{7}$ (c) $\frac{7}{12}$

   (d) $\frac{5}{12}$ (e) $\frac{11}{12}$

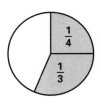

2. $(3ab^4)^2(-2a^2b^{-3})^4 =$

   (a) $-48a^{10}b^{-4}$ (b) $-48a^{16}b^{-96}$ (c) $144a^{18}b^{97}$

   (d) $144a^{10}b^{-4}$ (e) $-144a^{10}b^{-4}$

3. What is the slope of the line with equation $5x + 3(x - y) + y = 2$?

   (a) -4 (b) 4 (c) -1 (d) 8 (e) $\frac{1}{4}$

4. If $f(x)$ is a quadratic function with graph as shown, then $f(x) > 0$ whenever:

   (a) $x > 0$

   (b) $x > 2$

   (c) $x < 1$

   (d) $-1 < x < 2$

   (e) $x < -1$ or $x > 2$

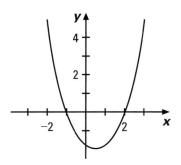

5. The x-coordinate of one of the points of intersection of the graphs of $y = x^2 + 2$ and $y = x + 4$ is

   (a) 0 (b) 6 (c) 2 (d) -2 (e) 3

6. If $3M = 6 - x$, for which values of x is $M \leq 15$?

(a) $x \geq -39$ (b) $x \leq -39$ (c) $x \geq -51$ (d) $x \geq -9$ (e) $x \leq -3$

7. Which of the following could be an equation for the graph below?

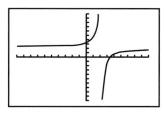

(a) $y = \dfrac{7 - x}{1 + x}$ (b) $y = \dfrac{7 + 3x}{2 - x}$ (c) $y = \dfrac{7 - 2x}{2 - x}$

(d) $y = \dfrac{x}{x - 2}$ (e) $y = \dfrac{3}{2 - x}$

8. The length of one side of a cube is x. If the length of each side is tripled, what is the volume of the new figure?

(a) $27x^3$ (b) x^3 (c) $9x$ (d) $9x^3$ (e) $18x^2$

9. For what values of x is $3^x \leq 0$?

(a) $x = 0$ (b) $x = 1$ (c) $x < 0$

(d) no value of x (e) $x \leq 0$

10. In the right triangle shown, $\cos \theta =$

(a) $\dfrac{4 - x}{4}$ (b) $1 - x$ (c) $\dfrac{\sqrt{16 - x^2}}{4}$

(d) $\dfrac{x}{4}$ (e) $\dfrac{4}{\sqrt{16 - x^2}}$

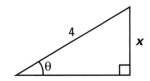

Lesson 2

Security—Cryptography

In the first lesson of this unit on informatics, you considered the issue of access. Certainly information is of little use if it is not organized and accessible. Another important issue is security. There are many situations where information must be secure, that is, private and authentic. For example, information security is important in business, particularly in e-commerce (the transaction of business on the Internet). For online bookstores, auction houses, and brokerages, it is essential that financial transactions are confidential and authentic.

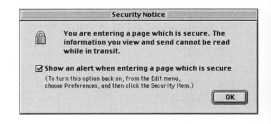

These same issues also arise with regard to the widespread use of electronic mail. You would like to know that the messages you send are read only by the intended recipients, and that the messages you receive are from who you think they are from. The mathematics used to keep information secure in all these situations is called *cryptography*.

COMPUTING
ZDNN TECH NEWS NOW

RSA weighs in on teen's 'breakthrough'
Crypto pros praise Irish high-school student, reserve judgment on new technique.

By *Robert Lemos*, ZDNN
January 18, 1999 9:00 PM PT

She's young, she's brilliant, and she knows what she's talking about. But to prove she's a real prodigy, her work must still endure the test of time.

Or so say encryption experts who gathered for the RSA Data Security Conference in San Jose, Calif., this week. They agreed that 16-year-old Sarah Flannery of Blarney, Ireland knows encryption, but they criticized media claims that her techniques are faster and better than established algorithms.

"She knows what she's talking about," said Ronald Rivest, Webster professor of electrical engineering and computer science at the Massachusetts Institute of Technology (MIT) and the 'R' in RSA. "But there's not enough information to evaluate her work."

NEWS BURSTS
March 12, 1999 AM

06:28a
BULLETIN: Motorola sues Intel
05:45a
Dow 10-K may be threat to computers
05:43a
Dole campaign wired for cash
04:43a
Microsoft to reorganize
04:25a
Stocks to watch

Think about the issue of information security as it relates to electronic mail, e-commerce, and other situations from your own life.

a E-mail is a common form of communication these days. Suppose you send a private e-mail message to a friend that you don't want anyone except your friend to read. How might you do this? How can your friend be sure the message is authentic and really came from you?

b Consider the security notification box on the previous page, which pops up at times when you are using a popular Internet browser. Why do you think this notification is given? What kind of information do you think needs to be kept secure? How do you suppose the browser software ensures that the "information you view and send cannot be read while in transit"?

c Refer to the article on the previous page about Sarah Flannery. What do you know about the terms "RSA" or "encryption"? The article refers to media claims that Ms. Flannery's techniques are faster than established algorithms. Why do you think a faster algorithm would be a "breakthrough" in cryptography?

INVESTIGATION 1 Symmetric-Key Cryptography

The design of methods for coding information makes extensive use of ideas and techniques from a branch of mathematics known as *cryptography*. **Cryptography** is the study of mathematical techniques used to provide information security. You will study two fundamental aspects of cryptography—*confidentiality* and *authentication*. Confidentiality is important when someone sends a credit card number over the Internet, or when a government official sends a secret message to an embassy abroad. Authentication refers to verifying the origin of data, which at the same time can verify the integrity of the data. Authentication is needed to ensure that a

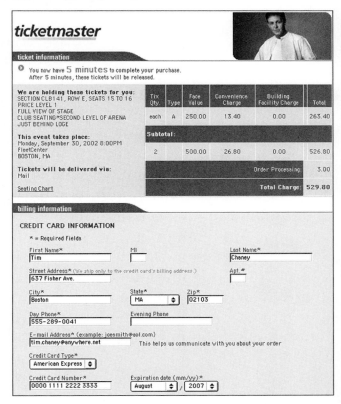

message or a contract sent by electronic mail has not been altered and is from who you think it's from.

Confidential information is transmitted and received as illustrated in the diagram below. The original plaintext message is encrypted. The resulting ciphertext message is decrypted upon receipt. A key is used to encrypt and decrypt the information. A **cryptosystem** is the overall method of encrypting and decrypting using keys.

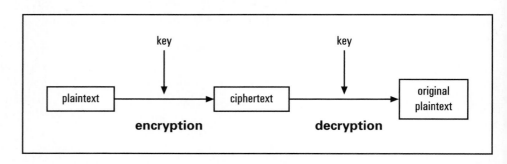

There are two basic types of cryptosystems. In a **symmetric-key cryptosystem**, the same key is used to encrypt and decrypt. (The exact same key might not be used, but at least it is easy to calculate the encryption key from the decryption key, and vice versa.) Thus, the security of a symmetric-key cryptosystem depends on the secrecy of the key.

In contrast, in a **public-key cryptosystem**, different keys are used for encryption and decryption, and the encryption key is made public. Thus, everyone knows how to encrypt messages, but only those who know the decryption key can decrypt the message. In this investigation, you will study symmetric-key cryptography. In Investigation 3, you will study public-key cryptography.

The following two symmetric-key cryptosystems replace letters from the plaintext message with letters that are a fixed number of places down the alphabet.

1. **ROT13** is a simple cryptosystem commonly found on UNIX-based computers and on e-mail programs. Using this system, every letter is replaced by the letter that is 13 places down the alphabet.

   a. Use this method to encrypt the message: VICTORY IS NEAR

   b. On a widely-used e-mail program, there is a feature that allows you to automatically encrypt messages using ROT13. For example, here is an e-mail message that was automatically encrypted:

      QRNE NZL,

      QBA'G GRYY NALBAR NOBHG GUVF!

      FRPERGYL LBHEF,

      JBBQL

   Decrypt this message.

   c. Explain why the same ROT13 procedure that encrypts a message will also decrypt it.

2. A fixed-shift letter-substitution cryptosystem is sometimes called a **Caesar cipher**. Reportedly, Julius Caesar (100–44 B.C.) used a method like this to send confidential messages. To decrypt a ciphertext message using a Caesar cipher, you need to know how many places down the alphabet you must shift to find the letter substitution (that is, you need to know the key). Suppose you have just intercepted the two one-line messages below that have been encrypted using the same Caesar cipher. Examine the ciphertext carefully to find the shift that has been used; then decrypt the two messages.

Julius Caesar used a fixed-shift letter-substitution cryptosystem.

JBBQ QLKFDEQ

F EXSB X AOBXJ

3. Because of the substitution technique used, the above methods are examples of *substitution ciphers*. A **cipher** is an algorithm used for encryption and decryption. (Generally, there are two related algorithms, one for encryption and one for decryption.) A **substitution cipher** is an algorithm that replaces each character in the plaintext with another character. The receiver reverses the substitution to decrypt the message. A **key** typically consists of a single number or several numbers that specifically determine the encryption (or decryption) process. For example, the key in a Caesar cipher is the number of places down the alphabet that you must shift.

a. What is the key for the Caesar cipher in Activity 2?

b. What is the key for the ROT13 cipher?

c. Substitution ciphers like the ones above are not very secure. That is, it is not very difficult to decrypt the ciphertext even if you don't know the key. Describe some general strategies you think could be used to break substitution ciphers like those above.

d. Create your own substitution cipher. Encrypt a brief message. Then exchange encrypted messages with a classmate and try to decrypt each others messages. Discuss and resolve any difficulties.

4. One reason that substitution ciphers are not difficult to break is that the letters of the alphabet used in the English language are used with different frequencies. For example, "e" is the most commonly used letter in the English language. Thus, by examining enough ciphertext, you could deduce which letter substitutes for "e," and this would tell you the substitution shift used in the cipher. One way to defeat this method of breaking the cipher is to scramble the substitution of letters in such a way that the same letter may get a different substitute letter in different parts of the message. One method that accomplishes this is a more complex substitution cipher using matrix multiplication. This method was invented by the mathematician Lester S. Hill in the 1920s. It is called a **Hill cipher**.

  a. A specific type of Hill cipher is described below. Follow and complete the steps of the algorithm to see how this Hill cipher works.

  ■ Find a 2 × 2 matrix with integer entries that has an inverse with integer entries. For example, if $A = \begin{bmatrix} 1 & 2 \\ 1 & 3 \end{bmatrix}$, find A^{-1}.

  ■ Put the plaintext message into a 2 × n matrix. Use a dash (–) for the space between words. For example, MEET TONIGHT becomes

$$\begin{bmatrix} M & E & - & O & I & H \\ E & T & T & N & G & T \end{bmatrix}.$$

  ■ The first step is to substitute numbers for letters in the matrix message.

– (space)	A	B	C	D	E	F	G	H	I	J	K	L	M
0	1	2	3	4	5	6	7	8	9	10	11	12	13

N	O	P	Q	R	S	T	U	V	W	X	Y	Z
14	15	16	17	18	19	20	21	22	23	24	25	26

Complete the following matrix so that it becomes the number-version of the MEET TONIGHT matrix above.

$$\begin{bmatrix} ?? & 5 & 0 & 15 & 9 & ?? \\ 5 & 20 & ?? & ?? & 7 & ?? \end{bmatrix}$$

  ■ The next step of encryption is to do a scrambled substitution using matrix multiplication. Multiply the numerical message matrix on the left by the matrix A. Carry out this multiplication and record the result.

  ■ Now you have numerical ciphertext. But it would be nice to have the ciphertext as a string of letters, since that is the form of the original message (and besides, you don't want to give away any hints about your cipher by showing numbers or matrices). Work with some classmates to figure out how to translate the numbers in the ciphertext matrix into letters. As you do this, remember that you have 27 characters, numbered from 0 to 26. So, for example, the number 28 would translate to "A". Compare your translation with that of other members of your class. Resolve any differences. Finally, write the string of letters that is the final ciphertext.

b. Now that you know how to encrypt a message using a Hill cipher, it's time to think about the method of decryption. Starting with the final ciphertext from Part a, work with some classmates to figure out how you could decrypt this ciphertext. Then do the decryption. Check to make sure you end up with the original plaintext: MEET TONIGHT.

c. Compare the Hill ciphertext for the message MEET TONIGHT to the Caesar ciphertext for the same message, from Activity 2. Which cipher do you think is more secure? Why?

d. Use the Hill cipher in Part a to encrypt a one-word message of your choice. Exchange ciphertext messages with a partner. Decrypt each other's ciphertext message. Discuss and resolve any difficulties.

Checkpoint

In this investigation, you studied symmetric-key cryptosystems. In particular, you investigated several substitution ciphers—the ROT13 cipher, Caesar ciphers, and Hill ciphers.

a The diagram at the beginning of this investigation illustrates the basic process of sending and receiving confidential information. Demonstrate each step of this process using the ROT13 cipher and a one-word message.

b In a symmetric-key cryptosystem, the same key is used to both encrypt and decrypt information, or it is easy to determine the encryption key from the decryption key (and vice versa). Explain why each of the three cryptosystems studied in this investigation is a symmetric-key cryptosystem.

c Explain why all three ciphers studied in this investigation are examples of substitution ciphers.

d The main issue considered in this lesson is security. Explain why the Caesar and ROT13 ciphers are not very secure. Explain why the Hill cipher is more secure than the Caesar or ROT13 ciphers.

Be prepared to share your thinking with the class.

The U.S. government's *Data Encryption Standard* (*DES*) is a symmetric-key cryptosystem that has been the most commonly used cryptosystem up to recent times. It was first approved by the federal government in 1977. Since then, the DES has been widely used for electronic funds transfer and for the protection of civilian satellite communications. However, there are growing concerns about its vulnerability. For example, in January 1999 a network of professional and amateur cryptographers needed just over 22 hours to decipher a secret message encrypted with the DES algorithm. Due to clear indications of the lack of security using DES, the National Institute of Standards and Technology (NIST) launched a program in 1998 to seek a new *Advanced Encryption Standard* (*AES*).

Groups of mathematicians around the world submitted their best symmetric-key cryptosystems. At the Third AES Candidate Conference held April 13–14, 2000 in New York, the five finalists for the AES were presented and discussed. The winner, an algorithm called Rijndael (pronounced "Rain Doll"), became the official AES algorithm in 2001. Rijndael uses much larger keys than DES and thus it is much more secure. The DES was secure for 20 years. The new AES symmetric-key cryptosystem should remain secure for at least that long.

On Your Own

Think about the advantages and disadvantages of ROT13, Caesar, and Hill ciphers as you complete the following tasks.

a. The following two ciphertext messages were encrypted using a Caesar cipher. By examining the ciphertext, find the substitution key that was used and decrypt the messages.

PSSO FIJSVI CSY PIET

WIMDI XLI HEC

b. Use the ROT13 cipher to encrypt this message: SINGING IN THE RAIN.

c. Use a Hill cipher based on the matrix $A = \begin{bmatrix} -4 & -3 \\ 9 & 7 \end{bmatrix}$ to encrypt this message: TO BE OR NOT TO BE. Decrypt the message to verify that you get the original plaintext.

INVESTIGATION 2 ▸ Modular Arithmetic

In this investigation, you will examine more carefully the idea of a modular arithmetic. In order to use the Hill cipher in Investigation 1, you had to convert all integers in a numerical ciphertext to integers between 0 and 26. This is an example of what is called *modular arithmetic*.

1. Think about the Hill cipher you worked on in Investigation 1, on pages 604–605.

   a. Describe the method you used to convert numbers larger than 26 into letters or a blank. Describe how you converted negative numbers into letters or a blank.

   b. The Hill cipher is based on a *mod 27 system*, since you were limited to the 27 integers: 0, 1, 2, 3, …, 26. It is possible to do arithmetic in this system. For example, $2 + 3 = 5$ mod 27, just as in regular arithmetic. The arithmetic becomes more interesting when the numbers get larger. Perform the following computations in mod 27. (Think of the three-line equal sign ($\equiv$) simply as a sign indicating the computation result for now; it will be more carefully explained in Activity 3.)

      ■ Explain why $25 + 9 \equiv 7$ mod 27.

      ■ $18 + 14 \equiv \underline{?}$ mod 27

      ■ $3 \times 25 \equiv \underline{?}$ mod 27

      ■ $7^2 \equiv \underline{?}$ mod 27

      ■ $-5 \equiv \underline{?}$ mod 27

   c. Compare your results and explanations with other members of your group. Resolve any differences.

2. Now consider some other modular arithmetic systems.

   a. Consider a mod 31 system. Do the same computations as in Part b of Activity 1, using mod 31.

   b. Consider mod 15. Would you say that 23 and 38 are "equivalent mod 15"? Why?

3. The idea of two integers being "equivalent mod *n*" is formally defined as follows.

   Two integers are **equivalent mod *n*** if and only if they have the same remainder upon division by *n*.

   A "three-line equal sign" is used to denote equivalence mod *n*.
   Thus, $a \equiv b$ mod *n* is read as, "*a* is equivalent to *b* mod *n*."

   a. Use the definition above to show that:

      ■ $23 \equiv 8$ mod 5

      ■ $76 \equiv 4$ mod 9

      ■ $-3 \equiv 24$ mod 27

   b. Find four integers that are equivalent to 2 mod 7.

4. To **reduce an integer mod *n*** means to replace the integer by its remainder upon division by *n*. For example, if you reduce 58 mod 7 you get 2, since dividing 58 by 7 leaves a remainder of 2.

   a. Reduce each of the integers below, using the indicated modular arithmetic system.

   - 48 mod 5
   - 397 mod 10
   - −24 mod 7

   b. What are all the possible results when you reduce integers mod 5? When you reduce integers mod 12? When you reduce integers mod 348? When you reduce integers mod *n*? Explain.

   c. Suppose two integers reduce mod *n* to the same number. Are these two integers equivalent mod *n*? Prove your answer.

5. When exploring the mathematics behind public-key cryptography in the next investigation, you will sometimes need to reduce some fairly large numbers mod *n*. In particular, you will need to reduce powers of numbers.

   a. Let's start small. Consider reducing 10^2 mod 6. Using properties of exponents and what you know about mod *n* arithmetic, give a plausible explanation for each step in the following three different reduction methods.

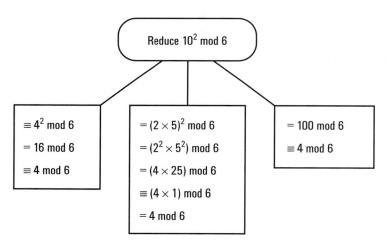

   b. Reduce 4^6 mod 12 in two different ways, as described below, to see if they each yield the same answer, and to see if there are any shortcuts.

   - Reduce 4^6 mod 12 by directly applying the definition of how to reduce mod 12. That is, compute 4^6, divide by 12, and find the remainder. What is the result?

- Using properties of exponents and what you know about mod n arithmetic, give an explanation for each step below:

$$4^6 = (4^2)^3 \bmod 12$$
$$= 16^3 \bmod 12$$
$$\equiv 4^3 \bmod 12$$
$$= (4^2 \times 4) \bmod 12$$
$$= (16 \times 4) \bmod 12$$
$$\equiv (4 \times 4) \bmod 12$$
$$= 16 \bmod 12$$
$$\equiv 4 \bmod 12$$

c. Based on Parts a and b you might conjecture the following:

Consider the arithmetic operations of addition, multiplication, and whole-number powers. When reducing mod n you can either perform these operations first and then reduce, or you can reduce first and then perform the operations. You get the same answer in either case.

Reduce 12^{25} mod 145. Compare your reducing strategy and answer to those of some of your classmates.

6. The cryptosystem that you will study in the next investigation uses a famous theorem first proved by the French mathematician Pierre de Fermat (1601–1665), called **Fermat's Little Theorem**:

If p is a prime number and a is any integer, then $a^p \equiv a \bmod p$.

Share the work among members of your class to check that this statement is true for several values of p and a.

Pierre de Fermat

In Activity 4 Part b, you found that every integer can be reduced mod 5 to 0, 1, 2, 3, or 4, since these are the possible remainders when you divide an integer by 5. You also found that, in general, every integer can be reduced mod n to an integer between 0 and $(n - 1)$, inclusive. This fact can be used to define a new set of numbers, called **integers mod n**:

$$Z_n = \{0, 1, 2, \ldots, n - 1\}.$$

Each "number" in Z_n really represents all the integers that reduce to that number mod n. Even so, you can think of the elements of Z_n as the numbers 0, 1, 2, …, $n - 1$. Arithmetic in Z_n is the same arithmetic mod n that you have been using in this investigation.

7. The modular arithmetic in Z_n has many interesting properties. Some properties are similar to properties of regular arithmetic with real numbers, while some properties are different. You will explore a few properties here; others will be considered in Investigation 3 and in the MORE set that follows it.

a. Recall that every real number x has an *additive inverse*, which when added to x yields 0. Find the additive inverse of each of these real numbers: 5, $\frac{3}{4}$, and -1.5.

b. Check to see if the additive inverse property is true in modular arithmetic systems.

- What is the additive inverse of 6 in Z_{10}?

- What is the additive inverse of 3 in Z_8?

- What is the additive inverse of m in Z_n?

Do you think every number in Z_n has an additive inverse? Explain your reasoning.

c. Recall that every nonzero real number x has a *multiplicative inverse*, which when multiplied by x yields 1. Find the multiplicative inverses of 5, $\frac{3}{4}$, and -1.5.

d. Check to see if the multiplicative inverse property is true in modular arithmetic systems.

- Consider the modular arithmetic system Z_7.

    i. Find a number in Z_7 that you can multiply by 3 to get 1 mod 7. Such a number is the multiplicative inverse of 3 in Z_7.

    ii. For each nonzero number in Z_7, try to find its multiplicative inverse.

- Now consider the modular arithmetic system Z_6.

    i. For each number in Z_6, try to find its multiplicative inverse.

    ii. State any patterns you notice concerning which numbers in Z_6 have a multiplicative inverse and which do not.

e. Explore possible conditions for multiplicative inverses to exist in Z_n. As you answer the questions below, be sure to consider several modular systems including Z_5, Z_6, Z_7, and Z_9.

- What must be true about the integer n if every nonzero integer in Z_n has a multiplicative inverse?

- For any modular arithmetic system Z_n and any nonzero element m in Z_n, how are m and n related if m has a multiplicative inverse?

You will be asked to prove your conjectures in Extending Task 2 (page 625).

In this investigation, you studied the modular arithmetic system Z_n and the associated operations of addition and multiplication mod n.

a The integers 8 and 5 can be considered elements in many different number systems. Think about the arithmetic of the various modular arithmetic systems as you complete the following:

- $8 + 5 =$ ___, in Z_9
- $8 + 5 =$ ___, in Z_{27}
- $8 \times 5 =$ ___, in Z_9
- $8 \times 5 =$ ___, in Z_{27}
- $8^5 =$ ___, in Z_9
- Find the additive inverse of 8 in Z_{12}.
- Find the multiplicative inverse of 5 in Z_{11}.

b Consider a mod 8 system.

- Describe how to determine if two integers are equivalent mod 8.
- Find three integers that are equivalent to -6 mod 8.
- Reduce 346 mod 8.

Be prepared to compare your responses to those of your classmates.

On Your Own

Carry out the indicated modular arithmetic computations.

a. $28 + 16 =$ ___, in Z_{30}

b. $37 + 25 \equiv$ ___ mod 7

c. Reduce 47 mod 12.

d. Reduce 12^8 mod 9.

INVESTIGATION 3 Public-Key Cryptography

In Investigation 1, you studied symmetric-key cryptosystems, in which the same key is used to encrypt and decrypt. For example, in the ROT13 system an alphabetic shift by 13 will encrypt and decrypt. In the Caesar cipher of that investigation, the alphabetic shift was 3 places. In a Hill cipher, the key is a matrix, which works to encrypt (multiply by the key matrix) and decrypt (multiply by the inverse of the key matrix).

One of the major security problems with symmetric-key cryptosystems is that the key must be confidentially distributed to the sender and receiver. This *key distribution problem* is eliminated in public-key cryptosystems.

In **public-key cryptosystems**, the receiver announces a public encrypting key that is known to everyone, but the receiver keeps secret a private decrypting key. Thus, anyone can encrypt a message and send it to the receiver, but only the receiver can decrypt the message.

The idea of public-key cryptosystems was developed in 1976 by Whitfield Diffie and Martin Hellman, two electrical engineers at Stanford University. In 1977, Ronald L. Rivest, Adi Shamir, and Leonard Adleman at the Massachusetts Institute of Technology developed a practical way to implement Diffie and Hellman's idea. Their method is now called the *RSA public-key cryptosystem*,

Stanford University campus, where the idea of public-key cryptosystems was developed

from the initials of their last names. Public-key cryptosystems are widely used today and are considered to be one of the most significant developments in the history of cryptography. In this investigation, you will study the RSA public-key cryptosystem.

The keys for the RSA public-key cryptosystem are numbers constructed by the receiver. The numbers are constructed using prime numbers and modular arithmetic. The general strategy is shown in the following diagram. The numbers *n* and *e* are used for encrypting. They comprise the **public key**, known to everyone. The numbers *p*, *q*, and *d* are used for decrypting. They comprise the **private key**, known only to the receiver. The arrows in the diagram show which numbers are related to each other. The statements at the top of the next page explain how to encrypt and decrypt messages.

RSA Cryptosystem

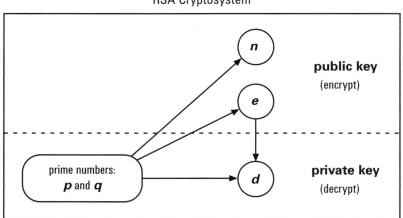

The Receiver Constructs the Keys:

- Choose two prime numbers, p and q.
- Compute $n = pq$.
- Compute $r = (p - 1)(q - 1)$.
- Choose a number e (for encrypt) such that e has a multiplicative inverse in Z_r.
- d (for decrypt) is the multiplicative inverse of e in Z_r.
- The receiver publishes e and n in a public directory. This is the receiver's public encryption key.
- The receiver keeps d secret, along with p and q. This is the private decryption key.

Encrypting:

- Convert the plaintext message, including spaces, to numbers in $Z_n = \{0, 1, \ldots, n - 1\}$.
- Raise each number to the power e. Reduce mod n.

Decrypting:

- Raise each ciphertext number to the power d. Reduce mod n.
- Convert from numbers to letters.

1. Encrypt, and then decrypt, the message FLEE using the RSA cryptosystem. For convenience, use the guidelines and tips given below.

   - First you need to construct the keys, that is, the numbers n, e, and d. To do that, you must choose two prime numbers, p and q. Normally you would use very large prime numbers to keep the system secure. However, for this activity, choose $p = 3$ and $q = 11$.

   - Convert letters to numbers by converting each letter into the integer 1–26 that represents its place in the alphabet.

   - Encrypt (and decrypt) one letter at a time.

   - Do not convert ciphertext numbers into ciphertext letters. You have finished encrypting when you have the ciphertext numbers.

2. Now, working with your teacher, form a team with at least one of your classmates and choose a name for your team. Your team should choose one of the public encryption keys listed in the public key directory on the next page. Enter your team name next to your public key on a copy of the Public Key Directory. Your teacher will hand each group their corresponding private key. Your team should use the RSA cryptosystem to encrypt and send at least one single-word message to one other team. Decrypt any messages you receive.

Public Key Directory		
Name	**n**	**e**
_____	85	13
_____	55	23
_____	161	19
_____	95	31
_____	91	29
_____	65	29
_____	35	13

Now that you know how the RSA cryptosystem works, consider two important questions: *Why does it work?* and *Why is it secure?* These questions will be the focus of the next few activities.

3. For this or any other cryptosystem to work, the decrypting process must undo the encrypting process so that you get back to the original message. In the case of the RSA system, you encrypt by raising to the power e (mod n), and you decrypt by raising to the power d (mod n), and this gets you back to the original message. So, for a number (message) M, $(M^e)^d \equiv M$ mod n. This works because of a special case of *Euler's Theorem*, which was proven by the Swiss mathematician Leonard Euler (1701–1783) using Fermat's Little Theorem (page 609).

   Euler's Theorem (special case): *($M^e)^d \equiv M$ mod n, where p and q are prime numbers, n = pq, r = (p – 1)(q – 1), and e and d are multiplicative inverses mod r.*

   a. Consider the following values:

   $$p = 3 \qquad q = 11 \qquad e = 7 \qquad d = 3$$

   - Verify that these numbers satisfy the conditions of Euler's Theorem.
   - Choose several values of M and verify that Euler's Theorem is true in this case.

   b. Choose two different (small) prime numbers and check that Euler's Theorem works for a few values of M.

4. Now you know how the RSA cryptosystem works and why it works. But why is it secure? The security of the RSA cryptosystem depends on the fact that multiplying two large prime numbers is easy, while factoring the product to recover the two prime numbers is difficult.

a. Look back at the diagram illustrating the RSA cryptosystem on page 612. Suppose you want to "crack the code." To do so, you need to find d, the private decryption key. Everyone knows n and e, since these comprise the public key. Suppose you could figure out what p and q are. Describe how you can find d if you know p, q, and e.

b. You saw in Part a that you can find d, and thus break the code, if you know p and q. But in a real application of RSA, p and q are secret. The only public information is e and n, the product of p and q. So now the question becomes, "Can you find p and q if you know the product of p and q?"

- Suppose $pq = 35$. What are p and q?

- Suppose $pq = 77$. Find p and q.

- Suppose $pq = 221$. Find p and q.

- Suppose $pq = 3{,}431$. Find p and q.

You can see that finding p and q gets more difficult as the product pq gets larger. In real applications of RSA, such as using a credit card to buy a book on-line, the key may include a product that contains 100 or more digits! The point is that it is very difficult to factor a large product into its prime factors. That's what keeps a secret message secret when you use public-key encryption.

A student browsing an online bookstore

5. Activity 4 illustrates a central idea in public-key cryptosystems—the notion of *one-way functions*. A **one-way function** is relatively easy to compute, but significantly harder to reverse. For example, multiplying two large prime numbers is a one-way function, since it's easy to do the multiplication but often quite hard to factor the product. One-way functions are important in public-key cryptosystems because encryption should be easy, while decryption without a key must be difficult. Of course, you don't want decryption to be difficult for everyone—the receiver must be able to easily decrypt. Thus, one-way functions by themselves are not useful as the basis for a cryptosystem. What you really need is a **trapdoor one-way function**, that is, a function that is difficult to reverse unless you have some special information (a "trapdoor"), like the key.

a. Consider the process of breaking a plate. Explain how this is like a one-way function.

b. Consider the process of putting mail into a public mailbox. Explain how this is like a trapdoor one-way function.

c. The encryption-decryption process used in the RSA public-key cryptosystem is an example of a trapdoor one-way function. There are three characteristics of a trapdoor one-way function, described below.

    i. It is easy to compute in one direction.

    ii. It is difficult to compute in the reverse direction.

    iii. If you know the secret, then it's easy to go in the reverse direction.

    How is each characteristic seen in the RSA system?

6. As you have seen, a good cryptosystem involves quite a lot of interesting mathematics. In addition, many technical details have to be worked out when a cryptosystem is implemented through specific software and hardware. Despite all these technical details, a cryptosystem is often initially designed or implemented by thinking in a non-technical way about *protocols*. A **protocol** is a series of steps, involving two or more parties (people or computers), designed to accomplish a task. For example, below is a protocol that describes how to implement a public-key cryptosystem. (It is conventional to use Alice and Bob to represent the two parties.)

> **Protocol for Implementing a Public-Key Cryptosystem**
>
>   **i.** Alice and Bob agree on a public-key cryptosystem, like RSA.
>
>   **ii.** Bob sends Alice his public key.
>
>   **iii.** Alice encrypts her message using Bob's public key and sends the ciphertext to Bob.
>
>   **iv.** Bob decrypts Alice's message using his (different) private key.

In the first investigation of this lesson, you studied symmetric-key cryptography. By carefully modifying the protocol above, write a protocol for implementing a symmetric-key cryptosystem.

7. Public-key and symmetric-key cryptosystems are not really competitors. Each is better suited for certain purposes since they each have different strengths and weaknesses. One way to compare the two types of cryptosystems is in terms of speed versus key security. Symmetric-key systems are faster, but less reliable in terms of key security. In particular, since the same key is used to encrypt and decrypt, one key must be sent to all authorized users. This makes it more difficult to keep the key secret and thus safeguard the whole system. Public-key systems are slower, but they work well for key distribution. The receiver closely guards his private decryption key, while the encryption key is freely transmitted as public knowledge. Given these complementary features of the two cryptosystems, sometimes a *hybrid cryptosystem* is used. This is described in the protocol on the next page.

Protocol for a Hybrid Cryptosystem

   i. Bob sends Alice his public key.

  ii. Alice generates a session key to be used for their communication session. She encrypts this session key using Bob's public key, and sends it to Bob.

 iii. Bob decrypts the session key using his private key.

  iv. Bob and Alice encrypt their communication using the same session key.

Describe where both symmetric-key and public-key cryptography are used in this protocol.

8. One of the strengths of public-key cryptography is that it can be used to authenticate data. Think about how you "authenticate data" with old-fashioned pen and paper. You use a signature. When you sign a check or a contract, that provides the certification that it is authentic. This same goal can be accom-

plished electronically using *digital signatures*. An effective digital signature can be created using public-key cryptography. Consider the following protocol.

Protocol for a Digital Signature Using Public-Key Cryptography

   i. Alice encrypts the document with her private key. This serves as her signature on the document.

  ii. Alice sends the signed document to Bob.

 iii. Bob decrypts the document with Alice's public key. This serves to verify the signature.

Suppose you receive an e-mail message from a friend that has been "signed" using the digital signature procedure above.

a. Explain how the digital signature verifies that the message is from your friend.

b. Explain how the digital signature also verifies that the message has not been altered.

In this investigation, you studied public-key cryptosystems. In the process, you learned about the RSA cryptosystem, trapdoor one-way functions, protocols, and digital signatures.

ⓐ Describe the processes of encryption and decryption used in the RSA cryptosystem.

ⓑ In the first investigation of this lesson, you studied symmetric-key cryptosystems. Describe some similarities and differences between symmetric-key and public-key cryptosystems.

ⓒ The main theme for this lesson is information security. Two important aspects of information security are confidentiality and authenticity. Explain how a public-key cryptosystem can be used to provide confidentiality. Explain how a public-key cryptosystem can be used to ensure authenticity.

ⓓ Look back at all three investigations of this lesson. List the mathematical concepts and techniques that you studied. For each, describe how it can be used to help provide information security.

Be prepared to share your descriptions and explanations with the entire class.

As you have learned in this lesson, the encryption process of public-key cryptography must be relatively easy to carry out. There are several steps involved in this process. Is each of the encryption steps easy? In fact, yes. First you need to generate some very large prime numbers, p and q. There are well-known computer algorithms that can quickly test very large numbers to see if they are prime. (For an example, see Organizing Task 5, page 622.) Running these algorithms on a good computer today, you can test whether a 100-digit number is prime in about half a minute. You already know that multiplying the two primes is relatively easy. Other crucial steps in the encryption process involve performing modular exponentiation and finding modular multiplicative inverses. These computations can also be done quickly using known algorithms and readily available computing technology. (See Extending Task 5, page 625.) Thus, encryption can be carried out relatively easily and quickly.

To ensure the security of the cryptosystem, decryption without a key must be prohibitively difficult. As you learned in this investigation, it all depends on how difficult it is to factor a large number that is the product of two very large primes. At the time of publication of this text, mathematicians have been unable to design a time-efficient algorithm for factoring large numbers. It is likely that factoring very large numbers will remain a difficult problem for a long time. But at the same time, new mathematical discoveries and more powerful computers make it necessary to use very large primes in order to ensure the security of public-key cryptosystems like RSA.

On Your Own

Use your understanding of the RSA public-key cryptosystem to complete the following tasks.

a. Suppose $p = 3$ and $q = 11$. Use these primes and the RSA cryptosystem to encrypt the one-letter message "D." Then decrypt to verify that you get back to "D."

b. There are at least three characteristics that a digital signature should have:

  i. The signature must be unforgeable, so that the receiver knows exactly who made the signature.

  ii. The signature must not be reusable, so that the receiver knows the signature was not cut from one document and pasted into a different document.

  iii. The signed document must be unalterable, so that the receiver knows the document was not changed after it was signed.

Explain how the digital signature defined by the protocol on page 617 has each of these characteristics.

MORE

Modeling • Organizing • Reflecting • Extending

Modeling

1. A variation on a Caesar cipher uses a keyword to shuffle the alphabet, and then you do a letter substitution. Here's how it works: First, choose a keyword that has no duplicate letters, such as, "crypto." Put this keyword in front of the standard alphabet, leaving out the letters in the alphabet that are in the keyword. Thus, in this case you get:

c, r, y, p, t, o, a, b, d, e, f, g, h, i, j, k, l, m, n, q, s, u, v, w, x, z

Now encrypt your plaintext by substituting letters based on this shifted and shuffled alphabet. For example, BE HAPPY becomes RT BCKKX.

a. Use this cryptosystem with the keyword "crypto" to encrypt the message MEET ME AT THE MOVIES. Decrypt the message NTIP BTGK.

b. Make up your own keyword and use it to encrypt a message.

c. Explain why this cipher is a substitution cipher but not a Caesar cipher.

2. Use a Hill cipher based on the matrix $A = \begin{bmatrix} 8 & -5 \\ 5 & -3 \end{bmatrix}$ to encrypt this message: BANANA SPLIT. Decrypt the message to verify that you get the original plaintext.

3. The basic coding technique of Caesar ciphers can be modeled using linear functions with arithmetic mod 26 and coding letters with A as 0 and Z as 25.

   a. Encode the message YOU HAVE MAIL in two ways:
      - Using the function $f(x) = x - 8$
      - Using the function $g(x) = x + 18$

      Explain the pattern relating the two resulting code sequences.

   b. The message EJDI OCZ XGPW was coded with the function $h(x) = x - 5$. What is the original message?

   c. If a message has been encoded using the function $C(x) = x + b$ with arithmetic mod 26, what general strategy will decode the message?

4. Suppose the following RSA public key is published: $n = 77$ and $e = 13$. You intercept a one-letter ciphertext message: "30."

   a. Can you break the code? What is the plaintext message?

   b. Why is it relatively easy to break this RSA code? Why are RSA codes in the real world impossible or at least very hard to break?

Organizing

1. Carry out the indicated modular arithmetic computations.

   a. $3 \times 4 = $ __ , in Z_5

   b. $18 + 36 \equiv x \bmod 7$, where x is in Z_7

   c. Find the multiplicative inverse of 8 in Z_{15}.

   d. Reduce $6^8 \bmod 14$.

2. The definition for *equivalent mod n* from Investigation 2 is the following:

   Two integers are **equivalent mod n** if and only if they have the same remainder upon division by n.

   Here is another definition that is often used:

   a and b are **equivalent mod n** if and only if $a - b$ is a multiple of n.

   That is, $a \equiv b \bmod n$ if and only if $a - b = mn$, where a, b, and m are integers and n is a positive integer.

   a. Verify that $46 \equiv 11 \bmod 7$ using both definitions.

   b. Prove that these two definitions are equivalent.

3. The addition and multiplication tables that arise from modular arithmetics have some very interesting properties.

  a. Complete the following addition and multiplication tables for arithmetic mod 7. These tables can be used to investigate properties of addition and multiplication in Z_7.

+	0	1	2	3	4	5	6
0	0	1					
1	1						
2	2						
3							
4				0			
5							
6							5

×	0	1	2	3	4	5	6
0	0	0	0				
1	0	1					
2	0						
3				2			
4						6	
5							
6							

  b. Perform these operations in Z_7.

   i. 6^2

   ii. $2 +$ (additive inverse of 6)

   iii. $4 \times$ (multiplicative inverse of 6)

   iv. $2 + (-3)$

  c. Solve the following equations in Z_7. Check your solutions.

   i. $3x + 6 = 2$

   ii. $x^2 + 3 = 5$

   iii. $(x + 4)(x - 6) = 0$

   iv. $3x + y = 5$ and $2x - y = 4$

  d. Which numbers in the system have *additive inverses*? Which have *multiplicative inverses*?

  e. Can you solve every linear equation of the form $ax + b = c$ when a, b, and c are integers, $a \neq 0$, in Z_7?

4. The Hill cipher method you used in Investigation 1 is not the most general version. In general, to use a Hill cipher you need an encrypting matrix that has integer entries and has a *determinant* that is *relatively prime* to n, where n is the number of letters in your alphabet plus other characters. The **determinant** of a 2×2 matrix $A = \begin{bmatrix} a & b \\ c & d \end{bmatrix}$ is denoted as $\det(A)$ and is defined to be $ad - bc$.

Two integers are **relatively prime** if their greatest common divisor is 1. Under these conditions the Hill cipher works as follows:

  a. First you need an encryption matrix. Find a 2×2 matrix E with integer entries such that $\det(E)$ is relatively prime to 27. (Also, to make it more interesting, make sure that $\det(E) \neq 1$.)

b. You encrypt a message using the same method as in Activity 4 on page 604. Using matrix E from Part a, encrypt the message: MATH. Write the final ciphertext.

c. The general decryption method is slightly modified from what you did before. Follow the procedure below.

■ Find the multiplicative inverse of det(E) mod 27.

■ Multiply E^{-1} by det(E). (Do you notice anything interesting when you do this?)

■ Multiply the resulting matrix by the multiplicative inverse of det(E). This gives you the decrypting matrix. Multiply by this matrix to decrypt the ciphertext in Part b. Verify that you get back to the original message: MATH.

5. An important part of the encryption process in the RSA cryptosystem is to find two large prime numbers. In this task, you will investigate how large numbers can be tested to see if they are prime.

a. List the first 10 prime numbers.

b. Determine whether or not 191 is prime. Describe the method you used.

c. The most tedious way to test a number N to see if it is prime is to start dividing N by all possible smaller positive integers.

■ Explain why you really only need to test-divide by *prime* numbers smaller than N.

■ Explain why you really only need to test-divide by prime numbers up to $\sqrt{N}$.

d. With large numbers, test-dividing by prime numbers up to $\sqrt{N}$ would require too many computations. Suppose N is a 50-digit number. Explain why $\sqrt{N}$, rounded to the nearest integer, is then a 25-digit number. There are many, many primes that are less than a 25-digit number! Test-dividing by all of them could take centuries even on the fastest computer. A better method is needed.

e. It turns out that Fermat's Little Theorem, from Investigation 2, can be used to test for primes. A slightly different form of this theorem is stated below:

If p is prime, and a is any number between 1 and $p - 1$ inclusive, then $a^{p-1} \equiv 1$ *mod* p.

In particular, if p is prime (and greater than 2), then $2^{p-1} \equiv 1$ mod p.

This gives a way to check a number N to see if it is *not* prime: Compute 2^{N-1} in mod N arithmetic. If you don't get 1, then N cannot be prime.

■ Compute $(2^9 - 1)$ mod 9. Do you get 1? Is 9 prime?

■ Try this test for $N = 15$.

Unfortunately, if you try this test and the result *does* come out to be 1, then you have no information. The number could be prime or not. For example, for $N = 341$, the result is 1. But 341 is not prime, since $341 = 11 \times 31$. So more work needs to be done to come up with a good prime tester. In fact, in 1986 a test was developed based on Fermat's Little Theorem that is a very good general-purpose prime tester. This method is called ARCLP, named after the five developers: Adleman, Rumely, Cohen, Lenstra, and Pomerance.

Reflecting

1. Symmetric-key cryptosystems are also called secret-key or private-key cryptosystems. Explain why the phrases "secret-key" and "private-key" are sensible descriptions of symmetric-key cryptosystems.

2. You can build a device like that shown below for quickly encrypting (or decrypting, if you know the key) messages using a Caesar cipher. The outer wheel represents the plaintext and the inner wheel represents the ciphertext. The inner wheel can be rotated to different positions.

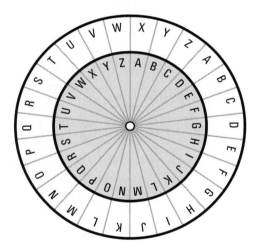

   a. Describe the particular cipher shown here in words and in an equation of the form $c = p + n \bmod 26$.

   b. How many different Caesar ciphers are possible? Explain your reasoning.

3. Mathematicians use the symbol Z_n to represent the finite set of numbers $\{0, 1, 2, 3, \ldots, n - 1\}$ because the letter Z is the first letter of the German word "zyklisch" which translates to "cyclic" in English. Why is the word "cyclic" appropriate in connection with modular arithmetic?

4. The news clipping at the beginning of this lesson is about a 16-year-old high school student in Ireland who made a "breakthrough" in cryptography. The article says that "her work must still endure the test of time." Do some research to find out whether her work has endured and write a short report.

5. Read the article about cryptography below.

U.S. to Lighten Up On Crypto Controls

by Jim Kerstetter

Encryption's Uncle Sam and Mr. Hyde?

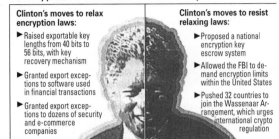

Clinton's moves to relax encryption laws:

► Raised exportable key lengths from 40 bits to 56 bits, with key recovery mechanism

► Granted export exceptions to software used in financial transactions

► Granted export exceptions to dozens of security and e-commerce companies

Clinton's moves to resist relaxing laws:

► Proposed a national encryption key escrow system

► Allowed the FBI to demand encryption limits within the United States

► Pushed 32 countries to join the Wassenaar Arrangement, which urges international crypto regulation

The Clinton administration is working on measures to relax encryption export laws for electronic commerce, despite ongoing concerns that the White House wants to toughen those regulations.

By month's end, the president's Export Council Subcommittee on Encryption is expected to release a new set of regulations regarding how encryption is used in e-commerce.

The regulations will define what can and cannot be encrypted in an online transaction, according to a spokeswoman for the Department of Commerce. The report, part of a wider administration effort to tackle the regulatory issues surrounding e-commerce, should allay fears that the Clinton administration is trying to hinder e-commerce merchants with tough encryption restrictions.

Source: *PC Week*, December 14, 1998, pages 1 and 16.

This article illustrates the many debates that have raged on how and if to regulate cryptosystems. For example, recall the Data Encryption Standard discussed on pages 605–606. Up until 1998, the full version of the DES, which had only a 56-bit key, was classified as a "munition," and export outside the United States and Canada was a felony. Thus, if you downloaded a strong version of, say, the Enigma cryptosystem (see Extending Task 1) from the Internet, and then sent it to a friend or business partner in Australia, you could be charged with a felony! In 1999 and 2000, this policy was relaxed, but still strong encryption algorithms with large keys were carefully regulated. Why do you think the full version of the DES was classified as a "munition"? Why do you think the U.S. government has been so concerned about exporting strong encryption algorithms?

Extending

1. There are several good cryptosystems that are available free on the Internet. For example, *Enigma* from Michael Watson and Next Wave Software (http://www.thenextwave.com/index.html) is a symmetric-key cryptosystem that implements the DES. Another freeware cryptosystem is *PGP—Pretty Good Privacy* (http://web.mit.edu/network/pgp.html), which was originally designed by Phil Zimmermann. At least one version of PGP uses RSA.

Find and download some cryptographic software from the Internet. Use the software to encrypt and decrypt some electronic files. Write a brief report describing the features and capabilities of the software.

2. In Activity 7 of Investigation 2 (page 610), you may have made the following conjectures:

   ■ For any modular arithmetic system Z_n and any nonzero element m in Z_n, m has a multiplicative inverse if the greatest common divisor of m and n is 1.

   ■ If n is prime, then every nonzero integer in Z_n has a multiplicative inverse.

   Prove each of these conjectures. In proving the first conjecture, you will need to use the fact that if the greatest common divisor of x and y is k; then you can find integers a and b so that $ax + by = k$.

3. To use a Hill cipher as in Investigation 1, you need an encrypting matrix E such that $E = \begin{bmatrix} a & b \\ c & d \end{bmatrix}$ and $E^{-1} = \begin{bmatrix} \frac{d}{ad-bc} & \frac{-b}{ad-bc} \\ \frac{-c}{ad-bc} & \frac{a}{ad-bc} \end{bmatrix}$ both have integer entries.

   a. Find such a 2 × 2 matrix (different from ones used so far in this lesson). Use this matrix to encrypt, and then decrypt, this message: MATH.

   b. The **determinant** of the 2 × 2 matrix E is denoted as det(E) and is defined to be $ad - bc$. It turns out that if both E and E^{-1} have only integer entries, then det(E) = 1 or −1. Verify this fact for the two Hill cipher matrices in Investigation 1 of this lesson: the matrix used in Activity 4 Part a (page 604) and the matrix used in the "On Your Own" task (page 606); and also for the matrix you constructed in Part a above.

   c. Prove the fact in Part b for 2 × 2 matrices.

4. At the end of Investigation 1, you read about the new Advanced Encryption Standard (AES), which has replaced the old Data Encryption Standard (DES). Write a brief report on either the DES or AES. Good resources include (a) "Standing the Test of Time: The Data Encryption Standard" by Susan Landau, in the March 2000 *Notices of the American Mathematical Society*, pp. 341–349 (Back issues of the *Notices* can be found on the American Mathematical Society Web site: http://www.ams.org/notices); (b) "Communications Security in the 21st Century: The Advanced Encryption Standard" by Susan Landau, in the April 2000 *Notices of the American Mathematical Society*, pp. 450–459; and (c) the AES Home Page at http://csrc.nist.gov/encryption/aes/.

5. An important part of the RSA cryptosystem is to find two numbers that are multiplicative inverses mod r. An efficient method for doing this involves what is called the *Euclidean algorithm*. Find a discrete mathematics or number theory book that describes this method for finding multiplicative inverses. Write a brief report on how the method works.

1. The ratio of water to concentrate in a juice mixture is 5 to 3. How many cups of water should be added to 15 cups of concentrate in order to make the juice?

   (a) 5   (b) 9   (c) 25   (d) 45   (e) 75

2. $\dfrac{5}{x} + \dfrac{2}{3x} =$

   (a) $\dfrac{17}{3x}$ (b) $\dfrac{7}{4x}$ (c) $\dfrac{17x}{3}$ (d) $\dfrac{10}{3x^2}$ (e) $\dfrac{7}{3x^2}$

3. In how many points do the graphs of $y = -2x - 3$ and $y = x^2$ intersect?

   (a) 0   (b) 1   (c) 2   (d) 3   (e) infinitely many

4. Using the standard coordinate system, which of the following is not the equation of a function?

   (a) $x^2 = y$ (b) $y^2 = x$ (c) $y = \sqrt{x}$ (d) $x = \dfrac{y}{2}$ (e) $y = \dfrac{5}{x}$

5. If $f(x) = x^2$, then $f(x - 3) =$

   (a) $x^2 - 9$ (b) $x^2 + 9$ (c) $x^2 + 6x - 9$

   (d) $x^2 - 6x - 9$ (e) $x^2 - 6x + 9$

6. Solve $(x + 5)(x - 4) = -8$.

   (a) $x = -4$ or $x = -13$ (b) $x = \pm\sqrt{12}$ (c) $x = -4$ or $x = 3$

   (d) $x = 4$ or $x = 13$ (e) $x = 4$ or $x = 3$

7. What is the domain of $y = \dfrac{x+4}{x^2-9}$?

    (a) All real numbers except -4

    (b) All real numbers except 9

    (c) All real numbers except 3

    (d) All real numbers except -3

    (e) All real numbers except 3 and -3

8. $\triangle ABC$ is similar to $\triangle ADE$. Find the value of x.

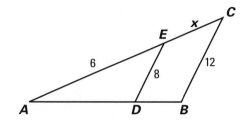

    (a) $x = 2$ **(b)** $x = 12$ **(c)** $x = 4$ **(d)** $x = 9$ **(e)** $x = 3$

9. Which of the following is equivalent to $\log x + 2 \log y - \log z$?

    (a) $\log \left(\dfrac{xy^2}{z} \right)$ **(b)** $\log (x + y^2 - z)$ **(c)** $\log (x + 2y - z)$

    (d) $2 \log \left(\dfrac{xy}{z} \right)$ **(e)** $\dfrac{\log x \cdot \log y^2}{\log z}$

10. Simplify: $\sqrt[3]{81} \cdot \sqrt[3]{18}$

    (a) $27\sqrt[3]{6}$ **(b)** $6\sqrt[3]{2}$ **(c)** $3\sqrt[3]{2}$ **(d)** $9\sqrt[3]{2}$ **(e)** $3\sqrt[6]{2}$

Accuracy—ID Numbers and Error-Detecting Codes

So far in this unit, you have learned how to make electronic information accessible and secure. It is also important for information to be accurate. When you play your favorite CD, you would like to hear the music as it was recorded, even if the CD is slightly scratched or you are playing it in your car on a bumpy road. When the cashier at the grocery store scans your food purchase, you expect that an accurate price will be sent to the cash register. When you check out a book from the library, it is important that the automated bar code reader sends the correct information to the library's database. However, anytime information is electronically read, recorded, or transmitted, there is a chance that errors could occur. In this lesson, you will learn some mathematical techniques for detecting and correcting such errors, primarily in the context of bar codes and ZIP codes.

Used by permission of Texas Instruments

49008-5201

Think About This Situation

Think about the issue of accuracy in the transmission of electronic information.

a What are some situations in your daily life where electronic information is read or transmitted and accuracy is important? How do you think accuracy is ensured?

b What kinds of information do you think are encoded on each of the three bar codes above? Why is accuracy of information important in these situations?

c What do ZIP codes of the home addresses of you and your classmates have in common? How are these common digits related to the issue of accuracy?

INVESTIGATION ▶ 1 Codes for ID Numbers: ZIP, UPC, and More

Identification (ID) numbers are found on books, driver's licenses, checks, airline tickets, mailing envelopes, grocery products, and in many other places. Today, many ID numbers are represented as bar codes. A **bar code** is a pattern of bars that encodes the ID number. Bar codes allow ID numbers to be scanned into computers, using laser scanning devices. Because of this feature, bar codes are sometimes referred to as "keyless data entry."

Obviously, an identification number should provide accurate identification. This is accomplished by using two types of digits in an ID number. Most of the digits, the **information digits**, are used to classify and identify the object. Then one or more extra digits are added to serve as *check digits*. The **check digits** are used to detect and correct errors that may occur when the number is entered into a computer or transmitted. For example, consider the U.S. Postal Service ZIP code.

ZIP codes (Zone Improvement Program codes) were first introduced in 1963 as a five-digit code. Nine-digit ZIP codes like the one on the business reply card below were introduced in 1983. The ZIP code identifies the postal delivery location. The first digit identifies a large geographical area, usually a group of states. The next two digits identify a particular mail-distribution center. The fourth and fifth digits indicate the town or local post office. The four extra digits added after the dash identify a specific location within a town.

In March 1993, eleven-digit ZIP codes were introduced. The additional two digits give the last two digits of the street address or box number. These final two digits are often not printed; they only appear in the bar code. NOTE: Since nine-digit ZIP codes are more common, only nine-digit ZIP codes are analyzed in this investigation. See Modeling Task 2 to learn more about the eleven-digit ZIP code.

1. By representing the ZIP code with a bar code, as is done at the bottom of the reply card on the previous page, computers are able to read the code and help process the mail. The ZIP-code bar code is created using the *Postnet code* in which the decimal digits 0–9 are converted to **binary digits (bits)**, 0s and 1s, so that a computer can more easily read the numbers.

   a. Each decimal digit 0 to 9 is represented by a five-bit string that contains exactly two 1s. This is called a *2-out-of-5* code. How many different five-bit strings are possible that contain exactly two 1s? List them all.

   b. Each of the 2-out-of-5 binary strings from Part a will represent one of the digits 0 to 9, according to the "dictionary" given below.

Postnet Binary Code Dictionary			
Decimal Digit	2-out-of-5 Binary Code	Decimal Digit	2-out-of-5 Binary Code
1	00011	6	01100
2	00101	7	10001
3	00110	8	10010
4	01001	9	10100
5	01010	0	11000

   Compare your list of 2-out-of-5 binary strings from Part a to the list given in the dictionary above. Resolve any differences.

2. Now examine the series of long and short black bars in the Postnet bar code at the bottom of the business reply card on page 629. Each long bar represents a 1 and each short bar represents a 0. The very first and last long bars are not used in the code; they are only used to define the beginning and end.

   a. What five-bit string is represented by the first five bars in the code? Is this a 2-out-of-5 string?

   b. Write all the five-bit strings represented by the bars. Separate each five-bit string with a space so that you can easily distinguish them.

   c. Use the Postnet Binary Code Dictionary in Activity 1 to translate the five-bit strings into decimal digits. Compare the result to the ZIP code printed in the address on the card. Why do you think they are different?

3. The tenth digit that you found in Activity 2 is the *check digit*. It is used to detect and correct errors that may occur when the ZIP code is read, recorded, or transmitted by a computer. In this activity, you will discover how the check digit works.

   a. Add up all nine digits in the ZIP code on the reply card, and also add the tenth check digit. Then reduce mod 10.

b. Repeat Part a for the three additional ZIP codes below, with indicated check digits.

- 47402-9961, with check digit 8
- 80323-4506, with check digit 9
- 02174-4131, with check digit 7

c. By examining the results in Parts a and b, state the rule for how the check digit works.

d. Suppose that a ZIP code is printed out by a computer as 55441-6733, with check digit 4. Do you think a printing error has been made? Explain.

e. It is good to know that an error has been detected, but it would be even better if the error could be *corrected*. Given that errors are rare, it is reasonable to assume that the error is probably in just one digit. Try to correct the error in the nine-digit ZIP code in Part d by changing one digit. Are there other one-digit changes you could make so that the check-digit rule is satisfied? Explain.

4. In Part e of Activity 3, you found that there were several possible one-digit changes that were consistent with the check-digit system. Since there is more than one way to "correct" the

detected error, with no clear reason to choose one over another, a computer handling the mail would simply set it aside to be looked at by a person. In general, the check digit by itself cannot be used to automatically correct errors. However, there are some errors that can be corrected if you use the check digit along with information about the 2-out-of-5 binary code. Consider what happens when an error occurs in just one bar of the bar code.

a. Suppose that the first 21 bars of a ZIP-code bar code are printed as below. (As always, the first bar is not part of the code, it only marks the beginning.)

$$ |.|..||.||....|||.|.. $$

Find the bar printing error. Explain why it is an error. Is there a unique, single-bar correction that can be made? If so, explain. If not, list all the decimal digits that could be the corrected digit.

b. Explain why a single-bar error will always be detected when using the 2-out-of-5 binary code. Explain why a two-bar error may or may not be detected.

c. In Part a, you were only given information about the first four digits of a ZIP code. In that case, you could detect the single-bar error, but you could not correct it. Now consider a case where you are given more information. Suppose that there is a single-bar error in the first 21 bars of a potential ZIP code, but there are no other errors. For example, the first 21 bars are printed as below.

I,,,III,I,,,,I,,,,II,

The last five correct digits of the ZIP code are: 8-3447, and the correct check digit is 5. Find and correct the single-bar error. What is the corrected ZIP code?

d. Explain why you can always correct single-bar errors, assuming that there are no other errors.

5. For every code, there must be a process for encoding and a process for decoding. In Activity 2, you worked through the decoding process for the Postnet code. You started with the bar code, then you decoded to get a series of 5-bit binary strings, then you used the dictionary to decode the 5-bit strings into decimal digits, and you ended up with the ZIP code. Now reverse that process. Consider this ZIP code: 22091-1593. Find the check digit, and then encode the ZIP code and check digit into a bar code.

6. ZIP codes use one of the simplest bar code and check digit systems. There are many other codes that work in a similar but more complicated manner. One of the most common such codes is the Universal Product Code (UPC).

The UPC was first used in 1974 to identify grocery items. Now, it is used for almost all retail goods. For example, the UPC from a can of pie cherries, shown above, is 0-41345-51718-1. A particular brand and style of fly fishing rod has UPC 0-43372-60092-8. The first digit identifies the general type of product. The "0" in both examples here identifies these items as nationally branded products. The next five digits identify the manufacturer, and the following five digits are assigned by the manufacturer to identify the specific product. The last digit is the check digit. On the next page, see how the UPC check digit works.

a. Consider the fishing rod UPC: 0-43372-60092-8. Carry out the following algorithm.

- Add the digits in positions 1, 3, 5, 7, 9, and 11.
- Triple this sum.
- Add this result to the sum of the remaining digits.
- Reduce mod 10. This final result should be 0.

b. Carry out the check-digit algorithm for the pie cherries UPC: 0-41345-51718-1. If your final result is not 0, reexamine your work and resolve any problems.

c. Suppose the pie-cherries UPC is transmitted and received with an error in the third position: 0-46345-51718-1.

- Will this error be detected by the check digit? Explain.
- Can this error be corrected based on information from the check digit? Explain.

7. You have seen that although both the ZIP code and the UPC code are able to detect single-digit substitution errors, they usually are not able to correct those errors. Now think about another common type of error—transposing two digits.

a. Suppose the ZIP code 48195-9822 is incorrectly read by a computer as 48159-9822 (the fourth and fifth digits are transposed).

- Will this error be detected based on information provided by the check digit? Explain why or why not.
- Are there any digit transposition errors that will be detected by the ZIP code check-digit system?

b. The UPC code for a pair of jeans is 1-15431-60401-4.

- Suppose that this is incorrectly scanned with the seventh and eighth digits transposed. Will the UPC check digit system detect this error?
- Suppose the jeans UPC is incorrectly read with the second and fourth transposed. Will this error be detected?
- Give as complete a description as you can of the types of transposition errors that can be detected by the UPC check digit system.

The two codes you studied in this investigation are used to provide accurate identification numbers. There are many other such codes. Some of those codes include bar code representations or other machine-readable features. They all provide some level of error detection and correction. For example, the ISBN code for books uses a more complicated system based on mod 11 arithmetic to provide standardized identification numbers for books published anywhere in the world. U.S. Postal Service money orders and many traveler's checks use an identification number code based on mod 9 arithmetic. FedEx®, UPS®, most airlines, and most rental car companies use a system based on mod 7. Major credit card companies use a complicated code based on mod 10, called *Codabar*, to issue credit card identification numbers. The Department of Defense, the car manufacturers, and the health industry use *Code 39*, based on mod 43. Several of these codes will be explored in the following MORE set.

The Department of Defense uses a code system based on mod 43.

On Your Own

Apply what you have learned about codes for ID numbers and error checking to complete the following tasks.

a. Find a business reply card that shows a nine-digit ZIP code and the bar code representation. Decode the bar code and verify that you get the ZIP code.

b. Find the check digit for this ZIP code: 33664-4162.

c. Although machine scanners are regularly used to read the UPC number from a bar code, sometimes a sales clerk needs to manually key in the number. Unfortunately, the digits can be distorted or otherwise difficult to read. The bar code and UPC number shown at the right are for a particular piece of luggage. The check digit is illegible. What is the correct check digit? (Note that the first digit is the 0 on the far left.)

d. Find a retail product that has a twelve-digit UPC code. Verify that the check digit given on the code is correct.

MORE
Modeling • Organizing • Reflecting • Extending

Modeling

1. Since 1969, an international identification number system has been used for books around the world. Each book is assigned a unique code—an International Standard Book Number (ISBN). These codes allow books to be more accurately identified and more effectively organized, especially with computerized recordkeeping. For example, consider ISBN 0-380-00832-7. A 0 or 1 in the first position indicates that a book was published in an English-speaking country. The next block of digits identifies the publisher. The third block is assigned by the publisher from among the block of numbers the Library of Congress gives the publisher and it identifies the specific book. These first three blocks of digits can be of different lengths, depending on the publisher of the book. The last digit is a check digit. Here's how the check digit works:

There are ten digits in an ISBN: *abcdefghij*. Compute $10a + 9b + 8c + 7d + 6e + 5f + 4g + 3h + 2i + j$. If the number is valid, this sum will be equivalent to 0 mod 11.

a. Verify that the ISBN given above is valid.

b. Verify that the ISBN of the book you are now reading is valid.

c. Consider this code: 0-88385-423-0. Is this a valid ISBN? Justify your answer. (See Extending Task 1 to explore the error-detecting and correcting capability of the ISBN code.)

2. In this lesson, you analyzed the typical nine-digit ZIP code. As mentioned on page 629, there is also an eleven-digit ZIP code that is sometimes used to identify the mailing location even more precisely. Also, the tenth and eleventh digits help sort the mail into the order in which it will be delivered by the carrier. The eleven digits are contained in the bar code. Only nine or five digits are actually printed. For example, below is a return envelope with a nine-digit ZIP code printed, but eleven digits in the bar code. Actually, there are twelve digits in the bar code, including the check digit.

KALAMAZOO LOAVES AND FISHES
913 E ALCOTT ST
KALAMAZOO MI 49001-3853

Serving the hungry of greater Kalamazoo

a. The check digit is chosen so that the sum of all twelve digits will be equivalent to 0 mod 10. Verify that this is the case for the bar code above.

b. Verify that the tenth and eleventh digits give the last two digits of the street address or box number.

c. Find a reply card or envelope that has eleven digits plus a check digit in the ZIP code bar code. Verify that the check digit and tenth and eleventh digits have the properties described in Parts a and b.

3. Each bank in the U.S. has an identification number. This ID number is usually printed at the bottom left of all checks from the bank. The first eight digits of the ID number are information digits, which identify the particular bank. The ninth digit is a check digit. Suppose the eight information digits are a_1, a_2, a_3, a_4, a_5, a_6, a_7, and a_8. The check digit is chosen so that it is the last digit of $7a_1 + 3a_2 + 9a_3 + 7a_4 + 3a_5 + 9a_6 + 7a_7 + 3a_8$.

a. A bank in Iowa has the following ID number: 073901877. Verify that this number satisfies the check-digit system described above.

b. Explain why choosing the check digit as described above is the same as choosing the check digit so that it is equivalent to the indicated sum mod 10.

c. Find a check from a U.S. bank, look at the bottom of the check to find the nine-digit bank ID number, and verify that the ID number satisfies the check-digit system described in this task.

Organizing

1. Suppose that a computer is reading or sending a code written in binary digits, that is, the only digits used are 0 and 1. Recall that such digits are called *bits*. Suppose that the probability of an error in one bit place is p, and assume that errors in different bit places are independent of each other. Suppose that you transmit n bits.

   a. What is the probability of no errors in any bit place?

   b. What is the probability of exactly one error?

   c. What is the expected number of errors?

   d. What kind of probability distribution describes this situation?

   e. The assumption of independence is crucial here, but independence of errors is not always a legitimate assumption. Consider the case of adding two numbers. Is the probability of an error in one decimal place independent of errors in other decimal places? Explain and give an example. (When the errors are not independent, they are said to come in *bursts*.)

2. The *Soundex* code was established by the National Archives to help locate and accurately identify names in old records such as birth, death, and marriage certificates and passenger lists for the boats that brought immigrants to the United States. This was primarily done to help establish a person's age for Social Security eligibility, since many states did not start recording births until after 1900. Using the Soundex code, surnames (family names) that have the same or similar sound get the same code, even if the

   spellings are slightly different. This is done because spelling and pronunciation were not as standardized in the past as they are now. Thus, family names of immigrants may have been spelled in various ways when entered into the records, especially if the names were first transcribed into English from another language, particularly one written in a different alphabet. Soundex codes begin with the first letter of the surname followed by a three-digit code that represents the first three remaining consonants in the surname, as outlined on the next page.

Soundex Code

Take the first letter as is.

Code the other letters as follows:

- Ignore A, E, I, O, U, H, W, and Y.

- Code each of the next three consonants (except H, W, and Y, which are ignored) using the numbers below. Ignore all other letters.

$$1 = B, P, F, V$$
$$2 = C, S, G, J, K, Q, X, Z$$
$$3 = D, T$$
$$4 = L$$
$$5 = M, N$$
$$6 = R$$

- If there are now three digits in the code, you are done. If there are less than three digits in the code, then add 0s at the end as needed to get three digits.

For example, the surname Hardy is coded as H630.

a. What is the Soundex code for Hart? For Romano? For Williams?

b. When doing historical research, one should consider the possibility that names with the same Soundex code could be the same name, because of non-standardized spelling or slightly different pronunciation at the time the names were recorded. Could the names Hart and Hardy have been the same, but just recorded as different?

c. Suppose that when searching the 1860 census records, you find J. Watkins in North Platte, Nebraska. Then on 1870 census records, you find John Watson in North Platte. Based on evidence from the Soundex code, is it possible that these are really the same person?

d. Examine the Soundex coding scheme and explain why you think it was designed the way it was.

3. The ZIP and UPC codes you have studied in this lesson involve check-digit systems that are based on sums or weighted sums. This generally creates a stronger error-detecting code than if sums were not used at all. However, not all codes use sums. For example, consider U.S. Postal Service money orders. Each U.S. Postal Service money order has a unique serial number, which consists of ten digits followed by a check digit. As usual, the check digit is included to help ensure accurate identification of the money order. The check digit is found by reducing the ten-digit number mod 9.

a. Suppose a money order serial number begins with the digits 3953988757. What is the check digit?

b. Verify that the eleven-digit number 69584712355 satisfies the check-digit criterion for a U.S. Postal Service money order.

c. If you take a correct eleven-digit serial number for a U.S. Postal Service money order and subtract twice the check digit, then the result is a multiple of 9. Explain why this is true.

d. Illustrate and explain why this check-digit system will not detect the error of transposing two digits in the first ten digits.

Reflecting

1. The Universal Product Code (UPC) has appeared on so many products for so long that you probably take it for granted. However, in its first year of use in 1974, it was quite controversial. Think of some plausible reasons why it would be controversial. Then do some research to find out why and write a brief report. One possible resource is the article "Bar Codes: Reading Between the Lines" by Ed Liebowitz, in the February 1999 issue of *Smithsonian* magazine (Vol. 29, No. 11, pages 130–146).

2. In this lesson, you explored the mathematical structure of ZIP codes and UPC codes. In general, a **code** is a group of symbols that represents information together with a set of rules for interpreting the symbols. In what sense can Roman numerals and musical scores be thought of as codes?

3. Did you find any of the applications of mathematics in this lesson particularly surprising or unexpected? Which ones? Why?

Extending

1. In Modeling Task 1, you considered the ISBN code. The check-digit system for this code will detect any single error or any transposition error. It will not in general correct all errors, although you can often at least narrow down the possible correct numbers.

   a. Consider the alleged ISBN from Part c of Modeling Task 1: 0-88385-423-0. If you have not already done so in Modeling Task 1, verify that there is an error.

   b. In fact, there is a single error in the alleged ISBN in Part a. Can this error be corrected by reasoning about the check-digit system? If so, correct the error. If not, explain why not.

   c. Prove that the ISBN check-digit code will detect all single-position errors, that is, errors in which exactly one digit is incorrect.

   d. The ISBN system will also detect all transposition errors. Prove this fact.

2. Consider the 2-out-of-5 code used for ZIP codes in Investigation 1. In Part b of Activity 4 on page 631, you explained why all single-bar errors can be detected when using the 2-out-of-5 code. It is always desirable to have the shortest code words possible, so it would be nice to use a four-digit code in this situation if possible. However, a four-digit code could not have the error-detection capability that the 2-out-of-5 code has. Explain why this is true.

1. The ratio of boys to girls in Mrs. Armstrong's fifth-hour math class is 3 to 5. If there are 15 girls in the class, how many people are in the class?

   (a) 12   (b) 24   (c) 27   (d) 9   (e) 30

2. $2(6 - 2(9 - 11)) =$

   (a) –2   (b) –68   (c) 20   (d) 8   (e) 16

3. Which of the following is an equation for the line containing (6, 4) and (4, –3)?

   (a) $2x - 7y = 34$ (b) $7y - 2x = 34$ (c) $7x + 2y = 34$

   (d) $7x - 2y = 34$ (e) $2x + 7y = 34$

4. If $f(x) = 3x^2 + 9x$ and $g(x) = 11x - 5$, then $f(x) + g(x) =$

   (a) $30x^2 + 20x + 5$ (b) $23x^3 + 5$ (c) $3x^2 + 20x - 5$

   (d) $28x^3$ (e) $3x^2 + 2x - 5$

5. One of the roots of $x^2 + x - 5 = 0$ is

   (a) $\dfrac{-1 + 2\sqrt{5}}{2}$ (b) $\dfrac{-1 - \sqrt{21}}{2}$ (c) $\dfrac{1 + \sqrt{21}}{2}$

   (d) $\dfrac{1 - 2\sqrt{5}}{2}$ (e) $1 + \sqrt{21}$

6. Solve for x: $9^{2x+1} = 27^x$

   (a) -2 (b) -1 (c) 2 (d) 0 (e) 1

7. If $f(x) = \frac{2x-1}{x+3}$, find the value of x when $f(x) = 0$.

   (a) $x = -3$ (b) $x = \frac{1}{2}$ (c) $x = -\frac{1}{2}$ (d) $x = -\frac{1}{3}$ (e) $x = \frac{1}{3}$

8. The perimeter of a rectangle is 56 meters, and its length is 4 meters more than its width. What is the area of the rectangle?

   (a) 192 m^2 (b) 780 m^2 (c) 16 m^2 (d) 182 m^2 (e) 144 m^2

9. If $\log_3 (x + 2) = 5$, then

   (a) $5^3 = x + 2$ (b) $3^5 = x + 2$ (c) $3(x + 2) = 5$

   (d) $x + 2 = 15$ (e) $(x + 2)^3 = 5$

10. The expression $8x^{\frac{2}{3}}$ can be written as

    (a) $\sqrt[3]{8x^2}$ (b) $2\sqrt[3]{2x^2}$ (c) $8\sqrt[3]{x^2}$ (d) $\sqrt{8x^3}$ (e) $8\sqrt{x^3}$

Lesson 4

Looking Back

In this unit, you have learned about an important new field of mathematics called informatics—the mathematics of information processing. Informatics is particularly important in our modern world of computers and the Internet. You investigated three themes of informatics: access, security, and accuracy. In Lesson 1 on access, you learned some of the mathematics behind Internet search engines, namely, set theory. In Lesson 2 on security, you studied cryptography and modular arithmetic. In Lesson 3 on accuracy, you learned about error-detecting codes used for ID numbers. In this final lesson, you will review and pull together all these ideas and apply them in new contexts.

1. In Lesson 3, you studied several different codes used for ID numbers. All those codes use a check digit and basic modular arithmetic. None are able to detect all single-position errors and all transposition errors without avoiding certain numbers or using other characters. (The ISBN code in Modeling Task 1 and Extending Task 1 of Lesson 3 detects these errors, but needs to use the extra character X.) In this task, you will investigate the *Verhoeff code*, which does not use modular arithmetic, but does detect all single-position and transposition errors. This code was used, for example, for the serial numbers on German banknotes. (Adapted from J.A. Gallian "Math on Money," *Math Horizons*, November 1995, 10–11.)

   a. The Verhoeff code was devised by the Dutch mathematician J. Verhoeff in 1969. It is based on what is called the *dihedral group* with ten elements, denoted D_{10}. The ten elements of D_{10} consist of five reflections and five rotations of a regular pentagon. Consider the regular pentagon *ABCDE* below.

   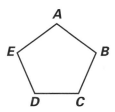

   - Explain why the regular pentagon can be rotated counterclockwise about its center through angles of 0°, 72°, 144°, 216°, and 288° and still coincide with itself. These are the five rotations in D_{10}.

   - Find five lines of reflection symmetry in the regular pentagon. The reflections across these five lines are the other five elements in D_{10}.

b. The first step in the Verhoeff code is to label the elements of D_{10} with the digits 0 through 9, as indicated below.

0: counterclockwise rotation of 0°

1: counterclockwise rotation of 72°

2: counterclockwise rotation of 144°

3: counterclockwise rotation of 216°

4: counterclockwise rotation of 288°

5–9: reflections across the lines of symmetry as shown in the diagram at the right

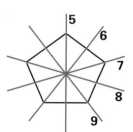

The elements in D_{10} are combined using function composition, denoted in this case by *, which can be interpreted as "followed by." Explain each of the following computations. Draw a diagram to illustrate each computation.

- $1 * 3 = 4$
- $2 * 8 = 5$
- $6 * 5 = 1$

c. Fill in the five missing entries in the table below, which shows how to combine all elements in D_{10}.

*	0	1	2	3	4	5	6	7	8	9
0	0	1	2	3	4	5	6	?	8	9
1	1	2	3	4	0	6	7	8	9	5
2	2	3	4	0	1	7	8	9	5	6
3	3	4	?	1	2	8	9	5	6	7
4	4	0	1	2	3	?	5	6	7	8
5	5	9	8	7	6	0	?	3	2	1
6	6	5	9	8	7	1	0	4	3	2
7	7	?	5	9	8	2	1	0	4	3
8	8	7	6	5	9	3	2	1	0	4
9	9	8	7	6	5	4	3	2	1	0

d. The Verhoeff code is similar to some other codes you have studied in that it uses a check-digit system based on a weighted sum. First you represent an ID number with the elements of D_{10}. Then you choose a check digit based on the outcome of a weighted sum. In a Verhoeff code, the "sum" is computed using the operation * from the table above, and the weights are assigned using the functions w_i that are defined in the table on the next page.

	0	1	2	3	4	5	6	7	8	9
w_1	1	5	7	6	**2**	8	3	0	9	4
w_2	5	8	0	3	7	9	6	1	4	2
w_3	8	9	1	6	0	4	3	5	2	7
w_4	9	4	5	3	1	2	6	8	7	0
w_5	4	2	8	6	5	7	3	9	0	1
w_6	2	7	9	3	8	0	6	4	1	5
w_7	7	0	4	6	9	1	3	2	5	8
w_8	0	1	2	3	4	5	6	7	8	9
w_9	1	5	7	6	2	8	3	0	9	4
w_{10}	5	8	0	3	7	9	6	1	4	2

For example, refer to the bold entry in the first row. This entry indicates that the element 4 in D_{10} would be weighted by the function w_1 to yield a value of 2. That is, $w_1(4) = 2$. Determine the following weights.

- $w_1(6)$

- $w_3(7)$

- $w_9(9)$

e. Now, here's how the Verhoeff code works: Suppose that the serial number on a German banknote is AU3630934N7. As usual, the last digit, 7 in this case, is a check digit used to ensure accuracy. Complete the following steps to see how the check digit is computed.

- Note that the banknote serial number includes some letters. Only ten letters were used in German banknote serial numbers, and they are converted to numbers according to the following table:

A	D	G	K	L	N	S	U	Y	Z
0	1	2	3	4	5	6	7	8	9

Convert the letters to numbers in the serial number given above.

- Thus, without the check digit, you now have 10 digits: $d_1, d_2, d_3, d_4, d_5, d_6, d_7, d_8, d_9,$ and d_{10}. Now compute a weighted sum by weighting each digit, d_i, with the corresponding function, w_i, and then "add" using the operation *. That is, compute the following "sum":

$$w_1(d_1) * w_2(d_2) * w_3(d_3) * \ldots * w_{10}(d_{10}).$$

- Finally, the check digit is chosen so that when it is also included (at the end, on the right), then the sum becomes 0. Verify that this is true for the given banknote number above and its check digit of 7.

f. Use the Verhoeff code to find the check digit for the German banknote serial number DA6819403G.

2. Complete the following tasks related to set theory and informatics.

    a. Suppose $S = \{2, 3, 9, 4, -2\}$ and $T = \{3, 6, 9, 12\}$.

       ■ Find $S \cup T$, $S \cap T$, and $S - T$.

       ■ Draw Venn diagrams illustrating the three sets you found above.

       ■ Let $V = \{x \mid x = 6n,$ where n is a positive integer$\}$. Find $T \cap V$.

    b. Suppose you are searching on the Internet for information about Republican candidates in Oregon running for either the House of Representatives or the Senate.

       ■ Represent this search using Boolean expressions.

       ■ Represent this search using sets.

    c. State whether each statement is true or false. If it is true, draw a diagram that illustrates the statement. If it is false, give a counterexample.

       ■ If $A \subseteq B$, then $A \cap B = A$.

       ■ If $3 \in S$, then $\{3\} \subseteq S$.

       ■ If $M \cap N = \varnothing$, then $M - N = M$.

       ■ $(A \cap B)' = A' \cap B'$

3. Complete the following tasks related to modular arithmetic.

    a. Reduce 12^3 mod 6.

    b. Compute $6 + 4$ and 6×4 in the following modular systems: mod 6, mod 8, and mod 20.

    c. Find three numbers that are equivalent to -3 mod 12.

    d. Solve $x + 7 = 2$ in mod 10.

    e. Show that in mod 6, the equation $x^2 - x = 0$ has four solutions. How can this be?

4. Use a Hill cipher based on the matrix $E = \begin{bmatrix} 5 & 2 \\ 7 & 3 \end{bmatrix}$ to encode, and then decode, the message SECRET.

5. Use the RSA cryptosystem with $p = 5$ and $q = 17$ to encode, and then decode, the message OK.

Checkpoint

In this unit, you have studied informatics—the mathematics of information processing. You have investigated three fundamental issues of informatics: access, security, and accuracy. These issues are reflected in the five main topics of the unit:

1. The mathematics of Internet search engines

2. Symmetric-key cryptosystems

3. Modular arithmetic

4. Public-key cryptosystems

5. Error-detecting codes for ID numbers

Working with your teacher, choose at least one of the five informatics topics listed above. Prepare a written and oral report on the topic chosen, according to the guidelines below.

- Give a brief overview and summary of the topic.
- Identify which of the three fundamental issues of informatics is addressed by the topic (access, security, or accuracy). Explain why and how this issue is addressed. Explain why this issue is important in information processing.
- List the specific mathematical methods and concepts related to your topic.
- Develop a thoroughly worked-out example illustrating the topic.

   Be prepared to present your group's report to the entire class.

On Your Own

Write, in outline form, a summary of the important mathematical concepts and methods developed in this unit. Organize your summary so that it can be used as a quick reference in your future work.

Problem Solving, Algorithms, and Spreadsheets

Money Adds Up and Multiplies

Every day you may see stories in newspapers and on television about wealthy people. Many of these people acquired their wealth through inheritance, starting successful businesses, and investments such as stocks and real estate, as well as from earnings in highly paid careers. A news story in the *Washington Post* suggests that it's not impossible for any one of us to become surprisingly wealthy and to do a great deal of good with that wealth.

Oseola McCarty was born in 1908 in rural Mississippi, and she had to quit school after the sixth grade to care for her aunt. Rather than returning to school, she started working. She started out doing laundry at $1.50 a bundle, but she also began setting aside part of her earnings in a savings account. By the time she retired 75 years later, she had accumulated a very substantial amount of money. She decided to give the majority of her savings for college scholarships to help students have access to the educational opportunities she never had for herself. In July 1995, the University of Southern Mississippi announced her gift of $150,000 to endow the Oseola McCarty Scholarship. Miss McCarty said, "I want young people to know that it's okay to save money, to invest it and have lots of money. More important, though, I want them to understand that what they don't need they should give to those who are less fortunate." Oseola McCarty received widespread recognition for her extraordinary gift, including an honorary doctoral degree from Harvard University, carrying the Olympic Torch, a visit to the White House, and many newspaper and television interviews. She passed away in 1999, but the scholarship fund will continue to help students for years to come.

Think About This Situation

The savings story of Oseola McCarty is very impressive. Even more admirable is her plan to give her savings to those who have been less fortunate.

a What pattern of savings do you think Miss McCarty followed in order to accumulate $150,000 for the scholarship fund after 75 years?

b If you wanted to have a $1,000,000 savings account 50 years from now, how much do you think you would have to invest now or over the years?

INVESTIGATION 1 Saving with Interest

Trying to figure out a savings plan that might lead to results like that of Oseola McCarty, or even more ambitious plans of your own, requires exploration of several key variables—the amount set aside each month or year, the interest rate that can be earned, and the length of time that the savings strategy is followed. One of the most useful mathematical tools to explore the relations among those variables is a computer *spreadsheet*.

A spreadsheet is an electronic matrix of cells in which numerical data or labels can be stored. More importantly, the cells of a spreadsheet can be related by formulas so that the numerical entry of one cell can be calculated from information in other cells according to rules that the spreadsheet user enters.

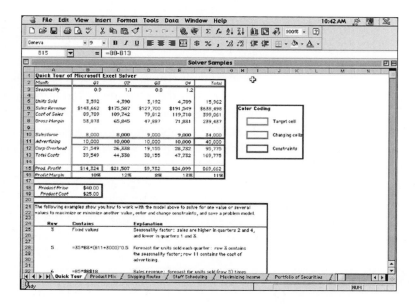

The following table shows a piece of one simple spreadsheet that tracks growth of money in a savings account built on an initial deposit of $25 and monthly deposits of $10 thereafter.

	A	B	C
1	Month	Bank Balance	Monthly Savings
2	Start=0	25	10
3	1	35	
4	2	45	
5	3	55	
6	4	65	

You could probably have figured out the savings plan displayed in the spreadsheet table above, but in the given form you can't see the power of the spreadsheet operation in producing such tables.

The table below displays the formulas used to calculate most of the entries in this spreadsheet. Compare cell entries in this spreadsheet display to the previous numerical display and see if you can figure out the use of symbolic expressions like "A3," "$C$2," and "=B3+$C$2" to guide calculations that produce the numerical values. Discuss your conclusions with classmates. Why do you think that $C$2 was used instead of 10?

B6	=B5+$C$2		
	A	**B**	**C**
1	Month	Bank Balance	Monthly Savings
2	Start=0	25	10
3	1	=B2+$C$2	
4	=A3+1	=B3+$C$2	
5	=A4+1	=B4+$C$2	
6	=A5+1	=B5+$C$2	

The next table shows the power of spreadsheets. You will notice that most numerical entries in the table are different from those in the original. However, all of those changes were caused by changing only one entry in the original spreadsheet. The number in cell C2 was changed from 10 to 15; then the formulas for cells B2 through B6 were used to recalculate the values in those cells.

	A	**B**	**C**
1	Month	Bank Balance	Monthly Savings
2	Start=0	25	15
3	1	40	
4	2	55	
5	3	70	
6	4	85	

The power of a spreadsheet is also enhanced by one more "ease of use" feature involving formulas. If you look back at the table showing formulas rather than numerical values in the cells, you will see simple repeated patterns in columns A and B. In column A each entry after row 3 is one more than the value in the preceding cell; in column B each entry after row 2 is obtained by adding the monthly savings amount (taken from cell C2) to the preceding cell value. The spreadsheet power comes from the fact that you only need to enter these *NOW-NEXT* formulas once. Then the spreadsheet command "fill down" will produce the formulas for succeeding cells automatically, changing the cell references as needed. If there is a cell value that you don't want to change as a formula is extended down a column, the spreadsheet convention involves using dollar signs ($) to fix the cell reference numbers, as in $C$2.

Most computer spreadsheets use a standard set of instructions for entering data and formulas. You will need to consult the manual or help features of the program that you are using for a few specific details. In a very short time, you will be creating spreadsheets that help you explore the interaction of many variables to solve problems.

1. The following part of a spreadsheet can be used to explore how Oseola McCarty could have saved $150,000 in 75 years. To limit the size of the spreadsheet, it is assumed that savings and interest on her bank account were deposited only once each year. This is only the start of the required spreadsheet.

	A	B	C
1	Year	Bank Balance	Yearly Savings
2	0	25	100
3	=A2+1	=(1+$C$4)*B2+$C$2	Interest Rate
4	=A3+1	=(1+$C$4)*B3+$C$2	0.05
5	=A4+1	=(1+$C$4)*B4+$C$2	
6	=A5+1	=(1+$C$4)*B5+$C$2	

a. Study the beginning of the spreadsheet to understand what each cell and formula represents. Then extend the sheet to cover 75 years. Does this choice of beginning balance, yearly savings, and interest rate result in $150,000 saved after 75 years?

b. Experiment with changes in the entries for initial deposit, yearly savings amount, and interest rate. Give four combinations of beginning balance, yearly savings, and interest rates that could produce $150,000 in 75 years.

c. Explain other factors that should be considered if the analysis is to reflect the probable pattern of Miss McCarty's savings and interest earnings, and show how the spreadsheet formulas could be adjusted to account for those factors.

2. For someone starting a savings plan today, it's quite reasonable to set a much higher goal than the $150,000 achieved by Oseola McCarty. Modify the spreadsheet of Activity 1 to explore combinations of initial deposits, annual savings, annual interest rates, and lengths of saving time that could make you a millionaire. Consider at least the following options:

■ Suppose first that the same amount of money is saved and invested each year.

■ Next consider the possibility that your increasing annual earnings would allow you to *increase* the annual savings amount in a steady pattern—either a constant dollar amount or a constant percent each year.

After you've explored various combinations of annual savings, interest rates, and lengths of saving time, write a report summarizing your findings.

3. For people who save throughout their working lifetimes, retirement means a change in income patterns. Instead of saving more money, retirees often spend some of their savings and share their wealth with children and charities. Suppose that you manage to save $500,000 before retirement. In this activity, you will determine how much of that nest egg could be spent or given away each year and still leave $100,000 to be distributed at your death as an inheritance for your children or a donation to a charity.

- Think about the variables involved in this situation and design a spreadsheet that accounts for the various factors.

- Use your spreadsheet to find combinations of spending and investment earning factors that will allow you to reach the specified goal and variations to reach other goals. Write a report summarizing your findings.

4. Some savings vehicles such as bonds, certificates of deposit (CDs), or treasury securities require only an initial deposit. The growth of that deposit over time is all due to interest earned, which may be paid annually or more often.

   Suppose that you have $1,000 to invest in such an account and want to know when your investment will double. Obviously, the doubling time depends on the interest rate.

   a. Design a spreadsheet that shows interest earned and the new balance of the investment account for each year. Then use that spreadsheet to find the doubling time (in years) for annual interest rates of 4%, 5%, 6%, ..., 12%.

   b. How would you describe the relation between interest rate and doubling time for an investment? Can you find an equation relating annual interest rate R and doubling time T that matches the pattern in your data?

   c. Bankers and investment analysts often use a rule of thumb for estimating doubling time of an investment. The *rule of 72* suggests that the doubling time will be approximately $T = 72 \div R$. How well does this rule match your results for interest rates from 4% to 12%?

   d. Summarize your findings.

On Your Own

Suppose that a ninth-grade student has a new job and wants to begin saving for college. Design a spreadsheet that can be used to track the value of the student's savings each month for four years, and find several combinations of initial deposit, monthly deposit, and interest rate that will lead to accumulated savings of $5,000 in four years.

INVESTIGATION 2 Borrowing and Lending

In the American economy, people borrow money for many kinds of major purchases—from cars and homes to furniture and home appliances. Credit cards are used to charge small and large purchases. There are two basic factors at work in any such transaction: the lender charges interest on the outstanding balance of the loan or credit card purchases, and the borrower must pay that interest and the balance to eventually pay off the loan. Most of those loans require a minimum monthly payment. In deciding on a repayment plan, the task is to find a combination of interest rate and monthly payments that will reduce the outstanding balance to zero in some specified amount of time.

1. Suppose that a person is considering borrowing $5,000 from a credit union to help purchase a used car. The credit union charges 7.5% annual percentage rate (APR) interest and the purchaser is to make monthly payments of $200.

   a. Explain how the $4,831.25 loan balance after one month is calculated.

   b. Design a spreadsheet that shows how the loan balance will change over time for specified interest rates and monthly payments.

   c. Extend the working spreadsheet to include cells showing the total amount of money required to repay the loan and the amount of that money that is interest on the loan.

   d. Use the spreadsheet to determine various combinations of interest rate, monthly payment, and loan length. Summarize your findings.

2. When young people enter college, they are often invited to open credit card accounts. What many students don't realize is that credit card interest rates can be very high, often as much as an 18% annual rate or more. Usually the companies require a minimum monthly payment of approximately 2% of the unpaid balance.

   Suppose that a first-year student uses his new credit card to charge a $2,500 stereo system and decides to make only the minimum payment each month. He also decides that he won't make any other purchases with the credit card until the stereo is paid off.

   a. Design a spreadsheet to determine the student's monthly credit card balance over a period of at least 5 years. The spreadsheet will be most informative if it includes columns showing the dollar values of the monthly interest charged, the monthly payment required, and the remaining balance.

   b. Find out how long it would take the student to pay off the debt on the credit card and the amount that would actually have been paid for the stereo system, including the amount for interest. To find these totals it will be helpful to use the spreadsheet SUM function. For example, to add the entries in cells C1 to C15, you can enter the formula "=SUM(C1:C15)" in cell C16.

   c. Explore ways that changes in the monthly payment and interest rate affect the total amount paid and time needed to pay off the charge. (Some banks charge lower credit card interest rates like 12% or 14%.) Summarize your findings.

3. Most people who buy houses or condominiums borrow part of the money needed for the purchase. These home mortgages can be for amounts as high as several hundred thousand dollars or even over a million dollars. Annual interest rates vary, but are usually between 6% and 12%, depending on general economic conditions in the country. Mortgage loans typically take as long as 30 years to repay. Consider a $75,000 home mortgage loan.

   a. Design a spreadsheet that will allow you to track the balance of such a loan under various interest rates and repayment amounts. To simplify things, you might consider interest and payments on an annual basis, rather than the more typical monthly procedure. Include columns that show the amount of each payment that goes toward interest and the amount that goes toward reducing the outstanding loan balance. Then define spreadsheet cells that give the total amount of interest and principal paid over the entire life of the mortgage.

   b. Use your spreadsheet to explore various combinations of interest rate, loan amount, and length of loan. Write up your results in a report that illustrates several plans that a home buyer might consider.

4. The interest rate on a credit card or loan is generally stated as an Annual Percentage Rate (APR). However, because interest is generally charged on a monthly basis, the *effective interest rate* can be higher. Suppose that you have a loan balance of $100 and make no payments for 12 months. Each month there will be interest charges added to the outstanding balance.

   a. Design a spreadsheet that shows the growth of this debt at the end of each month in a year. Include columns in the spreadsheet that give both interest charged and new balance for each month. Include separate cells that allow you to quickly experiment with different interest rates and initial balances.

   b. Find the balance at the end of one year for various interest rates from 4% to 20%. Use those balances to find the effective annual interest rate in each case.

   c. Repeat the experiments to find effective interest rates with a loan balance of $1,000 and compare the results to those for a $100 balance.

   d. Summarize your findings.

Investigations 1 and 2 required you to design spreadsheets and use some basic spreadsheet features to find combinations of interest payments or charges and savings or loan repayments in familiar economic situations.

a In designing a spreadsheet from scratch, how would you decide on column choices?

b Describe what you have learned about these tasks in spreadsheet design:

- Writing and extending formulas
- Using variable and fixed cell references in formulas
- Changing cell values to test effects of variables

c What kinds of mathematical relationships are especially useful in the spreadsheet analyses of saving and borrowing?

Be prepared to share your understanding of spreadsheet design with the class.

On Your Own

Look back at Activity 1, 2, or 3 from Investigation 2. Write another question about the situation which could be answered efficiently using a spreadsheet. Design and use the spreadsheet to answer your question.

Lesson 2

Building a Library of Algorithms

When mathematics is used to solve problems in science, engineering, government, business, or industry, the work often involves a pattern of repeated calculations with changing input data. For example, in statistical quality control, samples are repeatedly drawn from a production process and their measurements are compared to standards and prior

samples. When airlines make flight plans for various scheduled trips, they use vertex-edge graphs to model the possible paths from takeoff point to destination. Then they use systematic procedures to search among those paths for the trip plan that will give shortest flying time. When a financial company needs to design repayment plans for loans, they rely on algebraic formulas to find the combination of loan amount, interest rate, and monthly payments that will meet constraints.

The procedures of statistics, graph theory, and algebra, which involve taking given data and producing answers to well-defined questions, are examples of *algorithms*. Because there are some standard calculations involved in many different algorithms, many businesses create libraries of algorithms to which one can turn when solving an application problem. Algorithms can be written in specialized computer programming languages or in the language of spreadsheets.

Think About This Situation

Think about the work of scientists, engineers, businesses, or government offices that you know about and the mathematical calculations that might be involved in doing the work.

a What repeated calculations are often involved in the work of those situations?

b What mathematical ideas are required to design algorithms for the repeated calculations?

INVESTIGATION 1 Formula Algorithms

Many routine calculations used by businesses are algorithms based on formulas that calculate output from one or more given pieces of input data. For example, in many stores electronic sensing devices read bar codes to identify the items being purchased. Then the cash register converts those codes into unit prices and calculates the total cost of a purchase. Any change in prices requires change in the cash register program. In some restaurants there is a separate cash register key for each item on the menu.

1. Suppose that you are responsible for programming the cash registers at a local take-out restaurant that offers five kinds of sandwiches, french fries, a salad, soft drinks, coffee, and tea.

   a. Complete the following table to design a spreadsheet that would calculate the bill for any given order, including a 5% tax. Include sandwiches of your own choosing and unit prices that seem reasonable. Cells in column D will have formulas showing how to calculate the cost for each item from the item unit price and the number ordered. You should be able to enter only one such formula in cell D2 and then get the others with the "fill down" command.

   To help calculate the subtotal of the bill, before tax is added, you need to add the various item costs in column D. One easy way to do this is with the standard spreadsheet function SUM. Next to the cell "Bill Subtotal" you can enter the formula "=SUM(Dm:Dn)" where Dm is the cell label of the first item cost and Dn is the cell label of the last item cost.

   Before going ahead to the activities that follow, test your spreadsheet with some simple data to be sure it is working correctly.

	A	B	C	D
1	Item	Unit Price	No. Ordered	Item Cost
2	Hamburger	0.95	x	?
3				
4				
5				
⋮	⋮	⋮	⋮	⋮
			Bill Subtotal	
			Tax	
			Total Bill	

   b. Use the spreadsheet that you created to explore varieties of orders that produce a total bill of approximately $50.

   c. For any transaction at the restaurant, the cash register uses input data on the number of each item ordered. Assign single letter variable names for the numbers of each menu item in one customer's order (for example, H for hamburger, and so on). Then write an equation showing how the total bill B is calculated from the numbers of individual items ordered.

2. At the beginning of the twenty-first century, people are accustomed to the idea of satellites orbiting Earth for telecommunication and research purposes. One of the important problems in the launching of such satellites is determining the velocity required to place a satellite in the desired orbit.

■ The general formula relating velocity v (in kilometers per hour) and orbital radius r (in kilometers from the center of the primary body) is given by $v = \sqrt{\dfrac{GM}{r}}$.

■ G is a universal constant of gravitation and M is the mass of the primary body. For Earth, $GM \approx 5.08 \times 10^{12}$ km^3/hr^2.

■ The radius of Earth is about 6,370 km, so for a satellite in a circular orbit at an altitude of 1,000 km above the surface of Earth, the radius of that orbit is about 7,370 km.

a. The following table shows one plan for a spreadsheet that is helpful in exploring the connection between altitude and velocity of a satellite in orbit around Earth, our Moon, and the planet Jupiter. Use the given information about velocity and radii of Earth orbits to figure out the meaning of each cell in the spreadsheet. Complete the formulas in the columns for the Moon and Jupiter and find the orbital velocity for several different altitudes above each planet.

	A	B	C	D
1	Variable Names	Earth	Moon	Jupiter
2	?	5.08E+12	6.096E+10	1.615E+15
3	?	6,370	1,730	70,000
4	?			
5	?	=B3+B4	?	?
6	?	=SQRT(B2/B5)	?	?

b. Add a row **7** to your spreadsheet with the Variable Name "Orbit Length" and in columns **B**, **C**, and **D**, insert formulas that will give the length of orbits for any altitude. Then, add a row **8** with the Variable Name "Orbit Time" and in columns **B**, **C**, and **D**, insert formulas that will give the time in hours required for a satellite to make one complete orbit around the planet. Use the extended spreadsheet to estimate the orbital altitude, radius, and velocity for a satellite that takes exactly 24 hours to make one orbit of Earth. Such an orbit is called a *geosynchronous orbit* and makes the satellite appear to stay above one spot on Earth. These are the orbits used by current communication satellites.

c. The problem in Part b can also be solved by using algebraic calculations to solve the following equation for r: $\frac{2\pi r}{24} = \sqrt{\frac{5.08 \times 10^{12}}{r}}$. Solve this equation and use your result to verify the radius and altitude for the geosynchronous orbit estimated in Part b. Be prepared to explain why the given equation correctly models the problem.

d. How could you use algebraic reasoning to modify your spreadsheet to show how orbital radius or altitude depends on velocity, that is, so that the velocity cell could be changed and the matching radius and altitude calculated? Try your modified spreadsheet and compare results with those that you've produced in Parts a and b.

3. Every year there are millions of Americans who declare personal bankruptcy, often because they use credit cards to buy more than they can afford. Credit card companies require monthly payments and they charge high interest rates on the unpaid balance of an account. A rate of 1.5% per month is not uncommon. Suppose that someone builds up an unpaid credit card balance of $5,000, but then resolves to make no more new charges and to pay off the debt in full with equal monthly payments of $100.

a. Write a recursive formula of the form $y_n = ay_{n-1} - b$ with $y_0 = c$ that expresses the conditions of the credit card balance. What are the values of a, b, and c that match this problem situation? Use this recursive formula to find out how long it will take to pay off the debt at $100 per month.

b. For similar situations with monthly interest rate $r\%$ $\left(\text{so } a = 1 + \frac{r}{100}\right)$, there is a formula relating the unpaid balance y_n after n months to the values of a, b, y_0, and n. One version of the formula is $y_n = a^n y_0 - \frac{b}{a-1}(a^n - 1)$. Construct a spreadsheet that allows the user to select values of r, b, y_0, and n and then calculates a and y_n. Use this spreadsheet to approximate the monthly payment needed to pay off a $1,000 balance in 5 years (60 months) if the monthly interest rate is 1.5%.

- How will your answer change if only the interest rate is changed, reducing it to 1% per month?

- How will your answer change if only the time is changed, reducing it to 48 months (4 years)?

- How will your answer change if only the initial balance is changed, increasing it to $2,000?

In this investigation, you designed spreadsheets to analyze and solve problems using mathematical formulas.

ⓐ What factors about a problem situation need to be considered when deciding whether to use numbers or to use expressions with cell references in a spreadsheet design?

ⓑ How can cell entries be like mathematical formulas?

ⓒ Compare the spreadsheet formulas used in this investigation to the spreadsheet formulas in Lesson 1.

Be prepared to share your ideas with the entire class.

On Your Own

Recall that the formula for the standard deviation of a population is $\sigma = \sqrt{\frac{\sum (x - \bar{x})^2}{n}}$. Suppose you have a set of 100 data values. Provide a spreadsheet design that will calculate the standard deviation σ for this set of data values.

INVESTIGATION 2 Equation-Solving Algorithms

Problems that are modeled algebraically, including those that involve formulas of various types, often require solution of equations and inequalities. Thus, it is convenient to have algorithms available for routine solution of basic equation types. Then, all you have to do is enter the coefficients of the equations, and a tool like a spreadsheet will produce the solution automatically. In this investigation, you will develop several equation-solving algorithms.

1. Consider first linear equations and inequalities with one variable. Write spreadsheet instructions for the following general problem types. Then, test your designs with some specific equations and inequalities.

   a. Design a spreadsheet to solve linear equations in the form $ax + b = c$. You might base the layout of your spreadsheet on the following table, where the values of a, b, and c are to be entered in cells A3, B3, and C3 and the formula for the solution in D3.

	A	B	C	D
1	Solve:	ax+b=c		
2	a=	b=	c=	x=
3				

b. Design a spreadsheet to solve linear equations in the form $ax + b = cx + d$.

c. Design a spreadsheet to solve linear inequalities in the form $ax + b < c$.

d. Design a spreadsheet to solve linear inequalities in the form $ax + b < cx + d$.

One of the important things to keep in mind when designing spreadsheet algorithms for solving problems is to test the procedure you construct with some cases whose answers you already know. You should also try hard to consider all possible situations that might call for use of the algorithm you design and build formulas that work in all cases.

In testing your spreadsheets for the various questions in Activity 1, you might have encountered some complications in dealing with equations and inequalities like these:

$$3x + 1 = 3x + 2$$
$$-5x + 7 < 12$$

In both cases there are problem conditions that have to be checked before proceeding with arithmetic calculations. In the first case, you want to test whether $a = c$; in the second case you have to be careful with the negative coefficient of x.

Typical spreadsheet programs have a variety of logical-test functions that are useful in design of algorithms. For example, the formula A3=C3 tests equality of two cell entries. If the equality holds, then the spreadsheet reports that by a numerical value of "1". If the equality is false, the spreadsheet enters "0". There are also simple logical test functions for "<," ">," and a number of other statement types.

Running a test like A3=C3 is only the first step in dealing with situations where such a condition is critical. The next step is to tell the spreadsheet what to do in the cases of a true or false result. For this step, there is a logical operation that is fundamental to computer programming, the IF-THEN-ELSE instruction. In standard spreadsheet style, the IF-THEN-ELSE instruction is written in the following form:

=IF(logical test, do if test is true, do if test is false)

In the problem of solving equations like $ax + b = cx + d$, you could test for equality of the coefficients a and c with an IF-THEN-ELSE test like this:

=IF(A3=C3, "No Solution", (D3–B3)/(A3–C3))

To make an even more complete solution, covering the (pretty uninteresting) cases in which $a = c$ and $b = d$, you could use a nested IF-THEN-ELSE test like this:

=IF(A3=C3, IF(B3=D3, "All reals", "No Solution"), (D3–B3)/(A3–C3))

2. Modify your spreadsheet for solving equations like $ax + b = cx + d$ to include an IF-THEN-ELSE test. Check out the modification with several specific equations—some in which $a = c$, some in which $a \neq c$, and some in which both $a = c$ and $b = d$. Then, use a similar logical test function to modify your spreadsheet for solving linear inequalities. Be sure to check your modified spreadsheet in several cases for which you know solutions.

3. With algorithms for solving linear equations and inequalities available, the next natural step is designing an algorithm for solving quadratic equations of the form $ax^2 + bx + c = 0$, $a \neq 0$. Use what you know about solving such equations to build a spreadsheet algorithm that will deal with all possible cases. Be sure to test your algorithm on a wide variety of cases.

4. Many problems in arithmetic and geometry are answered by writing and solving proportions like $\frac{x}{a} = \frac{b}{c}$; $a, b, c \neq 0$. Use what you know about the solution of such equations to write a spreadsheet that will accept input of the values of three of the terms and solve for the remaining term.

5. Questions about exponential growth and decay quite often lead to algebraic equations in the form $b^x = c$; $b, c > 0$ and $b \neq 1$. Use what you know about logarithms to find a formula for solution of any such exponential equation. Build a spreadsheet that will use input values of b and c to produce the solution and then check the solution by calculating b^x. Test the spreadsheet with several cases whose solutions you know, such as $4^x = 64$, $1.5^x = 2.25$, or $2^x = 0.25$.

6. Modify your spreadsheet in Activity 5 so that it could be used to solve the general exponential equation $a(b^x) = c$, where $a, b, c > 0$ and $b \neq 1$.

Checkpoint

In this investigation, you designed spreadsheets for solving equations and inequalities.

a When is it helpful to use the IF-THEN-ELSE instruction in spreadsheet design?

b Think of another family of equations for which you could use the IF-THEN-ELSE instruction to write a solving algorithm.

Be prepared to compare your group's ideas to those of other groups.

On Your Own

Design and test a spreadsheet that will calculate the solution for absolute value inequalities of the form $|ax + b| \geq c$, $a \neq 0$. How would you modify your spreadsheet to calculate the solution for absolute value inequalities of the form $|ax + b| \leq c$?

1. $\frac{3}{4}\left(2 + \frac{1}{8}\right) - \frac{1}{2} =$

   (a) $\frac{35}{32}$ (b) $-\frac{7}{32}$ (c) $\frac{5}{3}$ (d) $-\frac{11}{40}$ (e) $\frac{4}{15}$

2. $6x - 4(x + y) - 2y =$

   (a) -4 (b) $2x + 6y$ (c) $-3xy$ (d) $2(x - 3y)$ (e) $2x + 2y$

3. The slope of the line $2x + 9y - 7 = 0$ is

   (a) $\frac{2}{9}$ (b) $\frac{9}{2}$ (c) $-\frac{2}{9}$ (d) $-\frac{9}{2}$ (e) $\frac{7}{9}$

4. If $f(x) = 2x^2 + 1$, then $f(a + 3) =$

   (a) $2a^2 + 10$ (b) $2a^2 + 19$ (c) $2a^2 + 12a + 18$

   (d) $2a^2 + 12a + 19$ (e) $2a^2 + 6a + 10$

5. One of the roots of $2x^2 + x - 6 = 0$ is

   (a) $-\frac{3}{2}$ (b) 2 (c) -1 (d) 3 (e) $\frac{3}{2}$

6. If $ax + b = c - 3x$, then $x =$

   (a) $\frac{c - b}{a + 3}$ (b) $\frac{c - b}{3a}$ (c) $\frac{b - c}{a - 3}$ (d) $\frac{b - c}{a + 3}$ (e) $-\frac{bc}{3a}$

7. Which of the following is not a zero of $p(x) = (x^2 - 3x - 18)(x^2 - 4)$?

   (a) $x = 2$ (b) $x = -2$ (c) $x = 6$ (d) $x = -6$ (e) $x = -3$

8. The measure, in radians, of $\angle AOB$ below is

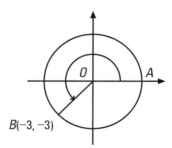

   (a) $\frac{\pi}{4}$ (b) $-\frac{\pi}{4}$ (c) $\frac{5\pi}{4}$ (d) $\frac{7\pi}{6}$ (e) $\frac{4\pi}{3}$

9. If $\log y = 3$, then $y =$

   (a) $1{,}000$ (b) $\frac{1}{1{,}000}$ (c) 10 (d) $\frac{3}{10}$ (e) $10{,}000$

10. If $\sqrt[5]{2x - 5} = 2$, then $x =$

   (a) 3.5 (b) 15.0 (c) 18.5 (d) 13.5 (e) 4.5

Lesson 3

Mathematical Patterns in Shapes and Numbers

The four polygons pictured below are stages in production of the geometric figure called the *Koch snowflake*. Starting with an equilateral triangle, the first step builds an equilateral triangle on the middle third of each side. That construction is repeated on each side of the new 12-sided figure, again on the resulting 48-sided figure, and so on.

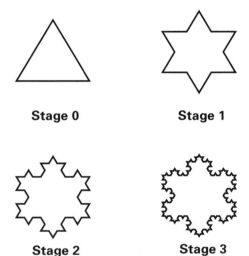

Stage 0   **Stage 1**

Stage 2   **Stage 3**

The Koch snowflake is a classic example of a **fractal**—a figure composed of parts that are similar to the shape of the whole. First created in 1904 by Swedish mathematician Helge von Koch, the snowflake is fascinating to many people with an eye for symmetry and design.

Think About This Situation

Imagine continuing the sequence of polygons in the creation of the Koch snowflake.

ⓐ How will the number of sides, the perimeter, and the area change from one stage to the next?

ⓑ Do you think there will be some upper limit to the perimeters of polygons in the sequence of polygons? To the areas? Explain your thinking.

INVESTIGATION 1 ▶ Fractals Forever

As you examined the progression of the number of sides, perimeter, and area in the stages of the Koch snowflake, you probably sensed some patterns that would allow you to predict the results at any future stage. To get actual numerical results, it helps to develop a recursive formula that will guide a calculator or spreadsheet to do the calculations for you.

1. Suppose that formation of a Koch snowflake begins with an equilateral triangle whose sides are each 1 inch long. Consider the number of sides, the length of each side, and the perimeter at each stage.

   a. Complete the following table by hand, showing the first several stages of fractal construction so that you can see how the patterns evolve.

   ### Koch Snowflake

Stage	Number of Sides	Side Length	Perimeter
0	3	1	3
1			
2			
3			

   b. Convert the table of Part a into a spreadsheet, using recursive formulas to generate entries in every cell except the starting values for the number of sides, side length, and perimeter.

   c. Look for patterns in the formulas of Part b and write function formulas for the number of sides, the length of each side, and the perimeter of the polygon at any stage n.

   d. Extend the spreadsheet to at least 30 rows and look for patterns in the numerical values that suggest answers to the following questions. Then reason from the symbolic formulas for side length and perimeter at stage n to justify your answers.

   - Is there a limit to the length of each side?

   - Is there a limit to the length of the perimeters of the polygons?

   e. How do entries in your spreadsheet change when the starting side length is increased or decreased? How is that result predictable from what you know about similarity of geometric figures?

2. Recall that the area of any equilateral triangle is a function of the side length.

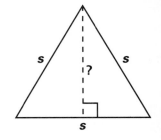

 a. Use the Pythagorean Theorem to express the area of an equilateral triangle in terms of its side length *s*. Write the formula in simplest radical form.

 b. Add an area column to your Koch snowflake spreadsheet and enter formulas in that column that will calculate the enclosed area at each stage of the snowflake development. Then use the spreadsheet to see if there is some limiting value for the enclosed area as the number of stages increases.

 c. How do the entries in the area column change as the length of a side of the starting equilateral triangle is increased or decreased? How is that result predictable from what you know about similarity?

 d. Some people find the relationships among side length, perimeter, and area of the Koch snowflake construction extremely puzzling, even "impossible." What is it about the patterns in those measurements that seems paradoxical?

The Koch snowflake is a two-dimensional figure, but some fractals are three-dimensional figures. One such fractal featured in the article, "Fractal Cards: A Space for Exploration in Geometry and Discrete Mathematics" by Elaine Simmt and Brent Davis, *The Mathematics Teacher* (February 1998, pp. 102–108), looks like this:

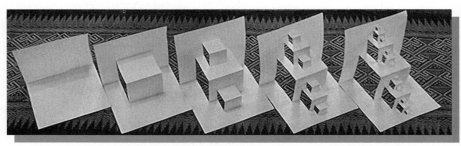

Stage 0   Stage 1   Stage 2   Stage 3   Stage 4

3. With your group members, make a model of stages 1 through 3 of this fractal by following these steps:

 ■ Start with a sheet of paper that is twice as long as it is wide. Fold this paper in half so that you have a square.

 ■ Then, along the folded edge, one-fourth of the way in from each side, make a cut perpendicular to the folded edge reaching halfway across the square.

 ■ Pop this section away from the folded edge and repeat the procedure on the two new shorter folded edges.

If your folding and cutting work is as intended, you'll be able to produce a pop-up figure that starts with one large cube and adds a sequence of progressively smaller cubical bumps.

a. Study the number and size of the cubes formed by the first several fold, cut, and pop-up operations. Organize data in a table like that which follows, assuming that the edge of the first cube has length 1.

Pop-up Fractal

Stage	Number of Cubes Added	Edge Length	Total Surface Area	Total Volume
1	1	1		
2				
3				
4				

b. Convert the table of Part a into a spreadsheet, using formulas to generate entries in every cell except the starting stage, number of cubes added, and edge length. Use the spreadsheet to extend the patterns to many stages and look for trends in the number of cubes added, edge length, total surface area, and total volume as the number of stages in the construction increases.

c. Use patterns in the spreadsheet data to find formulas that will allow you to make any of the calculations for the nth stage of construction without progressing through each prior step. Explain how those formulas confirm your conjectures about limiting values of edge length, surface area, and volume as the number of construction stages increases.

4. Design a spreadsheet to explore the *Sierpinski carpet*. Recall that the "carpet" is formed by cutting smaller and smaller squares out of the original square by always dividing each remaining square region into nine squares and removing the middle square.

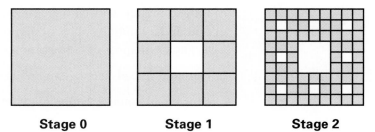

Stage 0 Stage 1 Stage 2

a. Set up a spreadsheet that allows you to study the trend in the area of the carpet as more and more regions are cut out. Explain how formulas used in the spreadsheet can be used to prove the observed trends.

b. As more and more regions are cut out from the original square, the resulting carpet has more and more exposed edges. Extend the spreadsheet from Part a to include the number of such edges added, the length of edges added, and the total length of exposed edges (perimeter) at each stage. Use that spreadsheet to look for a trend in the total length of exposed edges. Then explain how formulas used in the spreadsheet can be used to prove that observed pattern.

c. In what parts of the spreadsheets for Parts a and b did you use recursive formulas and in what parts did you use function formulas? What led you to each choice?

Checkpoint

Compare the results of investigating the Koch snowflake, the three-dimensional fractal, and the Sierpinski carpet.

ⓐ Summarize the patterns in the measurements. Describe surprises and similarities or differences that strike you.

ⓑ How did spreadsheets help you explore these patterns?

ⓒ How did reasoning with the symbolic formulas allow you to go beyond the spreadsheet-based observations?

Be prepared to compare your responses to those of other groups.

On Your Own

Draw the first three stages of the *Sierpinski triangle* formed by starting with an equilateral triangle and then removing the center equilateral triangles from the design at each stage.

a. Based on your work in this investigation, what patterns would you expect to find in the limiting values of area and total length of exposed edges (perimeter) as the number of construction stages increases?

b. Design a spreadsheet to explore these patterns and summarize your findings.

c. How could you use symbolic reasoning to prove the observed patterns?

Population Dynamics

The population of the world is now about 6.1 billion people. It is increasing at a rate of nearly a billion people every ten years. As you know, the world's population and its growth are not evenly distributed among countries or continents. For example, there are now about 1.2 billion people in more-developed countries and about 4.9 billion in less-developed countries. The growth rates of populations in those two sectors are also quite different—about 0.1% per year in more developed countries and about 1.6% per year in less-developed countries.

For governments trying to plan services like health care, housing, and education for their people, it's important to study the distributions of populations by age. The graph at the right shows the 2000 age profiles of two very different countries.

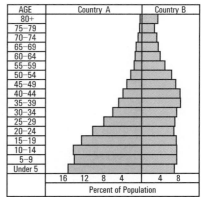

Source: www.census.gov/ipc/www/ idbsum.html

Think About This Situation

The science of *demography* is the statistical study of characteristics of human populations especially with reference to size, density, growth, distribution, and migration. It also studies factors that cause changes in those aspects of populations.

a The two countries for which population age profiles are given in the graph above are Kenya and the United States. Which portion of the graph do you think is the age profile for Kenya?

b How might the governments of these two countries use population age profiles when planning government services? What differences in government services might be the result of differences in the population age profiles?

c What factors might cause the observed differences in age profiles for the two countries?

d How could demographers make projections of change in age profiles of the two countries?

INVESTIGATION 1 ▶ Where Is the Population Headed?

Population data and projections can be given for many different political and social units—from cities and states to countries, regions, continents, and ethnic or religious groups. To track changes in populations there are three key factors—birth rates, death rates, and net migration rates (the difference between immigration and emigration rates). Spreadsheets can be very helpful in studying the effects of changes in these factors.

1. The following table gives basic demographic data, from the 2000 census, about the four major regions of the United States.

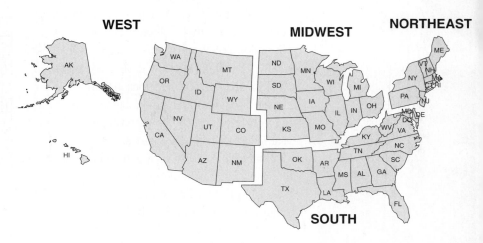

Region	2000 Population (in Millions)	Birth Rate	Death Rate	Net Migration Rate
Northeast	53.6	0.0128	0.0090	−0.0005
South	100.2	0.0149	0.0089	+0.0067
Midwest	64.4	0.0151	0.0090	−0.0011
West	63.2	0.0153	0.0070	+0.0078

Source: www.cdc.gov/nchs; www.census.gov

a. Construct a spreadsheet with the given information that will enable you to predict each region's population for the years from 2001 through 2020. Study the results of those predictions and make notes of interesting patterns.

b. Estimate the year in which the population of the South was expected to be twice that of the Northeast. Consider how this might have happened sooner with changes in migration rates.

c. Extend your spreadsheet to estimate when the population of the West will surpass that of the South and how changes in migration rates might affect this date.

d. Experiment with changes in the regional birth, death, and migration rates to see how changes in those factors affect populations.

2. The population distribution of the entire world has been changing fairly substantially over the past half-century, with much more rapid growth in certain parts of the world than in others. The following table gives some 2000 population figures for major continental regions of the world.

Region	2000 Population (in Millions)	Birth Rate	Death Rate
Africa	794	0.037	0.014
Asia	3,672	0.021	0.008
Europe	727	0.009	0.012
Latin America	519	0.022	0.007
North America	314	0.013	0.008
Oceania	31	0.017	0.008

Source: *World Population Prospects: The 2000 Revision*, United Nations, www.un.org

a. Modify the spreadsheet you constructed in Activity 1 to make projections of population change in regions of the world for each year from 2000 to 2015 or beyond, assuming that change factors stay as reported in 2000.

b. Many countries of the world have government policies to encourage or discourage population growth. They also work hard on improving public health to decrease infant mortality and to increase life span. Use your spreadsheet to explore the effects of changes in birth and death rates for the world as a whole.

In the "Think About This Situation" at the start of this lesson, you noticed that different countries can have very different age profiles. In less-developed and fast-growing countries, there are often many more children than older people.

3. In studying changes in the age profile of a country, the key factors are the current population at each age, the birth rates, and the "survival" rates from one age to the next. The following table gives such data on the United States from the 2000 census. Population figures are grouped, as is common practice, in 5-year intervals.

	5-Year Rates of Change	Age Intervals	Population (in Millions)	Population in 5 Years
Birth	0.071	< 5	18.9	$275.5 \times 0.071 \approx 19.6$
Survival	0.992	5–9	19.8	$18.9 \times 0.992 \approx 18.7$
	0.999	10–14	19.9	19.8
	0.999	15–19	19.9	19.9
	0.997	20–24	18.6	19.8
	0.995	25–29	17.9	
	0.995	30–34	19.6	
	0.994	35–39	22.3	
	0.992	40–44	22.6	
	0.988	45–49	19.9	
	0.982	50–54	17.3	
	0.974	55–59	13.3	
	0.960	60–64	10.7	
	0.938	65–69	9.4	
	0.906	70–74	8.8	
	0.859	75–79	7.4	
	0.792	> 80	9.2	
		Total	275.5	

Source: *National Vital Statistics Report*, Vol. 50, No. 6, March 21, 2002; www.census.gov/ipc/www/idbsum.html

a. Construct a spreadsheet that contains the information in the age profile table for the United States and allows you to extend the table to the right showing the profile for periods 5, 10, 15, 20, and 25 years in the future. It will be useful to keep some of these facts in mind:

- The five-year birth rate is a fraction of the total U.S. population.
- The five-year survival rates are the fraction of people in the preceding age interval who will be alive five years later (and be shown in the population figure for the next age interval). For example, 0.859 of the people in the 70–74 age group survive at least five years.

What population trends do you observe?

b. Revise your spreadsheet (saving the one from Part a) to improve the accuracy of the population model by using a different method of calculating births. The number of births in a country depends on the number of women in the childbearing years (roughly ages 15–45). In the United States, that leads to the following five-year rates in intervals from 15 to 49 years:

Age Interval	5-Year Birth Rate	% of Age Interval Female
15–19	0.2425	0.49
20–24	0.5615	0.46
25–29	0.6070	0.52
30–34	0.4705	0.49
35–39	0.2020	0.50
40–44	0.0395	0.50
45–49	0.0025	0.51

Source: www.censusscope.org/us/print_chart_age.html;
www.census.gov/ipc/www/idbsum.html

For example, if there are 20 million people ages 15–19, $0.49 \times 20 \times 0.2425$ or about 2.38 million births from that age group can be expected in five years. How do the population trends using the number of women in childbearing years compare to the trends observed in Part a?

4. As you saw in Activity 3, it looks as if the U.S. population will continue to grow. Use your spreadsheet from Part a of that activity to experiment with lower birth rates to find conditions that would lead to *zero population growth* in 25 years.

5. Most spreadsheets will allow you to construct histograms of data like the population age profiles in Activities 3 and 4.

   a. Use your spreadsheet software to make histograms of the age profiles of the U.S. population at the time of the 2000 census and 25 years later. Describe any changes that you feel are significant for American society and discuss some possible implications of those changes.

   b. Make another pair of histograms showing age profiles in 2000 and 25 years later for the conditions that would create zero population growth. Again, describe significant changes that are predicted by the graphs and discuss what those changes might mean for our society.

In this investigation, you used spreadsheets to make predictions of patterns in the growth of populations and to see how changing birth, death, and migration rates would affect those predictions.

a What instructions are needed with the spreadsheet software you use to produce each of the following:

- A cell formula that sums entries in a number of other cells?

- A histogram of entries in some column or row of a spreadsheet?

b What kinds of mathematics that you've studied in earlier work are most closely related to your spreadsheet models of population distribution and change?

Be prepared to explain your spreadsheet instructions and mathematical methods to the entire class.

On Your Own

In the "Think About This Situation" on page 671, you discussed the graph of the population age profiles of the United States and Kenya. The 2000 population of Kenya was about 30.4 million.

a. Modify the spreadsheets you have used to analyze U.S. population dynamics to study the situation in Kenya. Use the following estimates of the number of Kenyans in each age interval and the survival rates for each interval.

	5-Year Rates of Change	Age Intervals	Population (in Millions)
Birth	0.143	< 5	4.54
Survival	0.898	5–9	4.25
	0.991	10–14	4.20
	0.993	15–19	3.85
	0.982	20–24	3.23
	0.956	25–29	2.52
	0.917	30–34	1.91
	0.886	35–39	1.38
	0.889	40–44	1.06
	0.896	45–49	0.86
	0.904	50–54	0.70
	0.906	55–59	0.57
	0.904	60–64	0.45
	0.876	65–69	0.34
	0.826	70–74	0.24
	0.741	75–79	0.15
	0.623	> 80	0.11
		Total	30.36

Source: www3.who.int/whosis/life_tables/life_tables.xls

b. Make histograms of the age profiles for 2000 and 2025. Discuss possible implications of the population changes.

c. Explore the effects of change in the Kenyan birth rate on the total population and the population age profile for Kenya over several five-year periods into the future. See if you can find a *zero population growth* birth rate.

d. Explore the effects of public health care programs that might reduce infant mortality rates and other death rates at various ages. Report what you believe are some significant patterns in the numbers and their implications for Kenyan society.

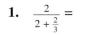

1. $\dfrac{2}{2+\frac{2}{3}} =$

   (a) $\dfrac{2}{3}$ (b) $\dfrac{3}{2}$ (c) $\dfrac{4}{3}$ (d) $\dfrac{3}{7}$ (e) $\dfrac{3}{4}$

2. Solve for x: $\dfrac{1}{x-2} + 3 = \dfrac{x}{x-2}$

   (a) $x = \dfrac{5}{2}$ (b) $x = 4$ (c) no solution   (d) $x = \dfrac{7}{4}$ (e) $x = -\dfrac{7}{2}$

3. Determine the value of y in the solution to this system of equations:

   $$x - 5y = -3$$
   $$2x + y = 5$$

   (a) $y = -1$ (b) $y = \dfrac{2}{11}$ (c) $y = 1$ (d) $y = 2$ (e) $y = -\dfrac{1}{11}$

4. If $f(x) = \dfrac{6-x}{x+6}$, then $f(9) =$

   (a) $\dfrac{1}{5}$ (b) -1 (c) 1 (d) -5 (e) $-\dfrac{1}{5}$

5. $x^2 - 9 > 0$ is equivalent to

   (a) $-3 < x < 3$ (b) $x < -3$ or $x > 3$ (c) $x > 3$

   (d) $x > \dfrac{9}{2}$ (e) $x < -\dfrac{9}{2}$ or $x > \dfrac{9}{2}$

6. Which of the following is equivalent to $4(5 - 2x) \le 15 - 3x$?

   (a) $x \ge -1$ **(b)** $x \ge 1$ **(c)** $x \le 1$ **(d)** $x \le -7$ **(e)** $x \ge -7$

7. If $f(x) = \dfrac{2x + 3}{4x + 1}$, then $f(a + 1) =$

   (a) $\dfrac{2a + 5}{4a + 5}$ **(b)** $\dfrac{1}{2}$ **(c)** $\dfrac{a + 4}{2a + 3}$ **(d)** $\dfrac{4}{3}$ **(e)** $\dfrac{2a + 4}{4a + 1}$

8. What is the period of the graph of $y = \cos 2x$?

   (a) $\dfrac{\pi}{4}$ **(b)** $\dfrac{\pi}{2}$ **(c)** 4π **(d)** 2π **(e)** π

9. $(6x^3y^{-2})^{-2} =$

   (a) $-12x^{-6}y^4$ **(b)** $\dfrac{y^4}{36x^6}$ **(c)** $\dfrac{y^4}{12x^6}$ **(d)** $\dfrac{1}{36x^6y^4}$ **(e)** $-\dfrac{12y^4}{x^6}$

10. $\dfrac{2x}{\sqrt[3]{7}} =$

   (a) $\dfrac{2x\sqrt[3]{49}}{7}$ **(b)** $\dfrac{2x\sqrt[3]{49}}{49}$ **(c)** $\dfrac{2x\sqrt[3]{7}}{7}$ **(d)** $\dfrac{2x\sqrt{7}}{7}$ **(e)** $\dfrac{\sqrt[3]{98x}}{7}$

Population and Political Power

One of the most important tasks for demographers is conducting *census* counts of populations to determine fair representation in government. The Constitution of the United States requires a complete national census every ten years. On the basis of the census results, there are often changes in the number of representatives to Congress from the various states.

In the United States Congress, every state has two senators, but the House of Representatives has 435 members who are elected by states in proportion to their populations. For example, in the 2000 census, California had an official population of 33,930,798. This was about 12% of the total U.S. population. California was allotted 53 representatives. There is a special requirement that each state must have at least one representative, regardless of its population.

Think About This Situation

As you can imagine, individual states are very interested in census results because they determine power in the federal government.

a Based on what you know of recent population trends, which states or regions are likely to gain representatives after the census in 2010? Which are likely to lose seats?

b Census results for individual states rarely lead to exact whole number results for representatives. For example, the exact calculation for California in 2000 suggests assigning 52.44715393 representatives. How do you think this problem should be dealt with in assigning congressional representatives to states?

c For many years, national censuses were conducted by sending Census Bureau staff members to every home address in the country. In recent years, those home visits have been replaced for many of us by mailed response forms. In planning for the year 2000 census, the Census Bureau proposed a technique that included projection of population totals from statistical samples, rather than collecting data from every household. Some politicians resisted this proposal. What groups of people do you think objected and why?

INVESTIGATION 1 ▶ Deciding Political Representation

The following table, on pages 681–682, gives some data about population and political representation in the 50 United States from the 2000 census, with projections for the year 2010. The states are grouped into geographic regions, and populations are rounded to the nearest 1,000 people.

State	2000 Population (in 1000s)	2000 Representatives	Projected 2010 Population (in 1000s)
Northeast			
New England			
Maine	1,278	2	1,323
New Hampshire	1,238	2	1,329
Vermont	610	1	651
Massachusetts	6,356	10	6,431
Rhode Island	1,050	2	1,038
Connecticut	3,410	5	3,400
Middle Atlantic			
New York	19,005	29	18,530
New Jersey	8,424	13	8,638
Pennsylvania	12,301	19	12,352
Midwest			
East North Central			
Ohio	11,375	18	11,505
Indiana	6,091	9	6,318
Illinois	12,439	19	12,515
Michigan	9,956	15	9,836
Wisconsin	5,371	8	5,590
West North Central			
Minnesota	4,926	8	5,147
Iowa	2,932	5	2,968
Missouri	5,606	9	5,864
North Dakota	644	1	690
South Dakota	757	1	826
Nebraska	1,715	3	1,806
Kansas	2,694	4	2,849

State	2000 Population (in 1000s)	2000 Representatives	Projected 2010 Population (in 1000s)
South			
South Atlantic			
Delaware	785	1	817
Maryland	5,308	8	5,657
Virginia	7,101	11	7,627
West Virginia	1,813	3	1,851
North Carolina	8,068	13	8,552
South Carolina	4,025	6	4,205
Georgia	8,207	13	8,824
Florida	16,029	25	17,363
East South Central			
Kentucky	4,049	6	4,170
Tennessee	5,700	9	6,180
Alabama	4,461	7	4,798
Mississippi	2,853	4	2,974
West South Central			
Arkansas	2,680	4	2,840
Louisiana	4,480	7	4,683
Oklahoma	3,459	5	3,639
Texas	20,904	32	22,857
West			
Mountain			
Montana	905	1	1,040
Idaho	1,297	2	1,557
Wyoming	495	1	607
Colorado	4,312	7	4,658
New Mexico	1,824	3	2,155
Arizona	5,141	8	5,522
Utah	2,237	3	2,551
Nevada	2,002	3	2,131
Pacific			
Washington	5,909	9	6,658
Oregon	3,429	5	3,803
California	33,931	53	37,644
Alaska	629	1	745
Hawaii	1,217	2	1,440
Total	**281,428**	**435**	**297,154**

Source: www.census.gov; *Congressional Apportionment: Census 2000 Brief*, July 2001

1. By entering these data into a suitable spreadsheet, you can explore different methods for allotting representatives to states and the way that population trends will lead to changes in national political power.

   First create a basic spreadsheet containing the population and representation data by state according to the 2000 census. To help in sorting the state data by population and representation, it will be convenient to put number codes for states and regions from 1 (for Northeast) to 63 (for Hawaii) in column A. Then the state names can go in column B, the 2000 census figures (in 1,000s) in column C, the number of representatives according to the 2000 census in column D, and the projected 2010 population in column E. Save this spreadsheet so that you can make many variations without reentering the data.

   a. Now focus on the question of how the number of representatives from each state is determined. In the basic U.S. population spreadsheet, delete the column E data on year 2010 population estimates. Replace those entries with formulas calculating the numbers of representatives to which each state would be entitled on a strict proportional basis, including fractional parts to two-decimal-place accuracy.

   b. Compare these "exact" figures to the actual whole numbers in use during the 2000s, and see if you can determine how the allotment of representatives to states rounds fractional parts to whole numbers.

2. Next do some calculations that might help to project changes in representation after the year 2010 census. Begin with the original spreadsheet that includes year 2010 population estimates in column E. Add two new columns to show the projected growth rates for each state from 2000 to 2010. Put the growth in number of people in the first of these new columns and the percent growth in the next column.

   a. Based on projections of growth in number of people, what states seem most likely to gain representation in Congress as a result of the 2010 census? Which seem most likely to lose representation?

   b. How do the percent growth rates give a different picture than the growth rates in numbers of people? Is the state with greatest percent growth rate the most likely to gain a representative? Is the state with lowest percent growth rate most likely to lose a representative?

   c. Which of the growth trends shown in the table are most consistent with things you have heard about population change in our country and which are surprising?

As you explored the relationship between state populations and numbers of representatives in the United States Congress, you probably discovered a variety of complications caused by the fractions that arise from proportional calculations. For example, it appears that in 2000 North Carolina was entitled to 12.47 representatives and Utah to 3.46 representatives. Normal rounding rules would suggest decreasing each to the nearest whole number: 12 for North Carolina and 3 for Utah. However, North Carolina actually got 13 and Utah got 3. If you applied standard rounding rules, you also found that a total of 433 representatives would be required.

3. Since change in the number of representatives from a state can mean a loss of political power for a state or loss of a job for someone in office, there have been many different proposals on how to deal with those fractions. The standard starting point in all schemes is the *people-per-representative* rate. Based on results from the 2000 census, that figure (in 1,000s) is $\frac{281,428}{435} \approx 646.96$.

   To find the number of representatives for each state, you can then simply divide the state population by the people-per-representative rate. For example, California, with 33,931,000 people would be entitled to $33,931 \div 646.96 = 52.45$ representatives. The question is: "What should be done with the fractions?" Brainstorm with classmates on possible ways to "deal with" the fractions in assigning representatives.

4. Alexander Hamilton, Secretary of the Treasury during the presidency of George Washington, proposed the following procedure (adapted to the current 435-seat House):

   ■ Calculate the exact number of representatives to which each state is entitled, including fractional parts.

   ■ Round the exact calculations *down* to the nearest lower integer value.

   ■ Add the 50 integer values and see how far short of 435 the total is.

   ■ Add single representatives to the states with highest fractional parts in the exact calculation stage until a total of 435 is achieved.

The "Hamilton method" (or "Vinton method") was used from 1850 to 1900. Use your basic spreadsheet created from the 2000 census data to test Hamilton's method. You can do the rounding with cell formulas like F2:=INT(E2). You can try to scan the fractional parts to see which are larger, or you can create a column containing only the fractional parts and then ask the spreadsheet to order those values from largest to smallest. The commands for ordering data in columns or rows are generally listed on the Data menu of a spreadsheet with the key word SORT.

Alexander Hamilton

a. How do the assigned numbers of representatives under Hamilton's method compare with the actual numbers of representatives allotted after the 2000 census?

b. What advantages and disadvantages can you see for Hamilton's method?

5. Thomas Jefferson, the third President of the United States, proposed a different procedure for assigning numbers of representatives in case of fractional results in the exact proportional calculations. The "Jefferson method" was used from 1790 to 1830. This method (adjusted to current conditions) is as follows:

■ As Hamilton suggested, calculate exact proportional values and round them down to the nearest integer values.

■ Then, instead of adding single representatives to selected states, revise the *people-per-representative* rate of 646.96 until the total of whole numbers of representatives sums to the desired 435.

Thomas Jefferson

To experiment to find a suitable rate, it might help to put the people-per-representative ratio in cell **F2**, and then refer to it with fixed reference in the formulas of column **E**.

a. Begin with **F2** as 645 and **E4** as the formula =INT(C4/$F$2) and then fill down column **E**. Keep an eye on the integer total at the bottom of column **E** as you experiment with different values of **F2**. Keep a record of any values of **F2** that give 435 representatives.

b. Compare the results of using Jefferson's procedure with the actual assignment of representatives used after the 2000 census and with the values obtained by use of Hamilton's procedure. For which states do the methods give different numbers of representatives?

c. What advantages or disadvantages can you see for Jefferson's method?

6. John Quincy Adams, another early United States president, proposed a mirror image of Jefferson's procedure. Instead of rounding down at the start, he proposed rounding up first and then adjusting the people-per-representative rate. (This method was never put into use.)

a. Explain why the formula E4:=INT(C4/$F$2)+1 will round the entry in that cell up to the nearest integer.

John Quincy Adams

b. Apply Adams' method until a suitable distribution of 435 representatives is achieved. Then compare the results and the people-per-representative rate to those produced by Jefferson's method.

7. Construct a table showing the numbers of representatives allotted by each of the three procedures—Hamilton's, Jefferson's, and Adams's—and see if you can find any patterns in the data suggesting which would be preferred by large states and which by small states.

8. The actual method of apportionment used by the United States Congress today was devised by mathematicians Edward Huntington and Joseph Hill. It is based on the **geometric mean** of two numbers a and b, which is the square root of their product. The Huntington-Hill strategy works as follows:

- Calculate the proportional representation for each state and round those values both down and up to the nearest integer values.
- Calculate the geometric mean of the two rounded values.
- If the proportional representation for a state is less than the geometric mean, choose the lower integer value; if it is greater than the geometric mean, choose the higher integer value.
- Adjust the people-per-representative rate until a total of 435 representatives is obtained.

For example, in 2000 the exact proportional representation for Colorado would have been $4,312 \div 646.96 = 6.67$. This rounds down to 6 and up to 7. The geometric mean of those two numbers is 6.48. This is less than the proportional share of 6.67, so Colorado was assigned the higher number of representatives, 7.

a. Put the proportional representation for each state in column E. Then use a spreadsheet formula like =SQRT(INT(E2)*(INT(E2)+1)) to find the geometric mean required for each state in the Huntington-Hill method. Enter those values in corresponding cells of column F.

b. Use a spreadsheet decision statement like =IF(E2<F2,INT(E2),INT(E2)+1) to select the lower integer value if the exact proportion is less than the geometric mean, or else the higher integer value if the exact proportion is greater than the geometric mean. Enter these decisions in column G.

c. Adjust the people-per-representative rate used for column E calculations until the sum of assigned representatives in column G is 435.

d. Compare the apportionment of representatives by the Huntington-Hill method to those of the Hamilton, Jefferson, and Adams methods and to the actual values given in the earlier data table. See if there is any pattern to the differences.

9. Now, return to the basic spreadsheet with a column containing the projected populations for each state in 2010. Use the Huntington-Hill method to find the numbers of representatives to which the states will be entitled if the new population census produces the projected results. Then write a short report in the form of a news article commenting on the big gainers and big losers in the adjustment of political influence in the U.S. Congress.

Checkpoint

In this lesson, you used spreadsheets to study a variety of methods for apportionment of political power in the United States Congress. In the process, you learned some additional spreadsheet functions and logical decision statements.

a What instructions are needed with the spreadsheet software you use to produce each of the following?

- Cell formulas that round values up or down to the nearest integer
- Cell formulas whose entries involve square roots of other cell entries
- Cell formulas whose entries are determined by IF-THEN-ELSE decisions based on other cell entries

b How did the spreadsheet technique of varying one cell entry in order to change the entire spreadsheet help as you explored options of the apportionment models?

c What spreadsheet instructions are needed to order the rows of a spreadsheet in ascending or descending order according to the numbers in one of the columns?

d What kinds of mathematics that you previously studied are most closely related to the spreadsheet models of apportionment?

Be prepared to share your group's spreadsheet methods and thinking with the class.

On Your Own

Consider the state in which you live and suppose that populations of every other state turn out as predicted for the year 2010. Experiment with your spreadsheet to see what increase in your state's population would be required to increase the number of representatives by 1. What state would lose that representative? Summarize your findings.

Lesson 6 · Looking Back

Spreadsheets are a versatile tool for modeling and analyzing problem situations. As you worked through the unit investigations, you reviewed many important mathematical ideas. You also strengthened your skills in algebraic reasoning. The activities in this final lesson will help you to consolidate your understandings of both algebraic and spreadsheet methods.

1. Five band students performed solo selections at the New Haven Music Competition. They were rated from 0 to 5 in two categories: technical merit and level of difficulty. The table below gives each student's ratings.

	Technical Merit	Level of Difficulty
Marie	4	3
Tony	4	4
Suji	5	3
Russ	3	4
Denise	3	5

Design a spreadsheet that will find overall ratings for each of the conditions below.

a. Technical merit and level of difficulty are rated equally important.

b. Level of difficulty is rated twice as important as technical merit.

c. Technical merit is rated twice as important as level of difficulty.

d. Add a third rating category of your choice, assign hypothetical ratings to each performer, and then find the competition results for your scenario.

2. Connections between and among mathematical ideas are sometimes surprising and curious. In previous mathematics study, you may have explored the Fibonacci sequence: 1, 1, 2, 3, 5, 8, 13, 21, 34, 55, 89, 144, 233, 377, 610, 987, ... This sequence has intrigued mathematicians for centuries. It appeared in the manuscript *Liber Abaci*, written in 1202 by Leonardo of Pisa, who was also known as Fibonacci.

Fibonacci

a. Enter 1 in cell **A1** and 1 in cell **A2** of a spreadsheet. Then use a recursive formula to generate 40 or more additional terms of the Fibonacci sequence in that column.

b. In column **B**, enter the ratios of successive terms in the Fibonacci sequence: $\frac{y_{n+1}}{y_n}$.

c. Study column **B** to see if you find a long-term trend in the entries.

d. In geometry, the *golden ratio* is the number that compares length and width of rectangles judged to be most attractive from an artistic or architectural point of view. The golden ratio is $\frac{1+\sqrt{5}}{2}$. What connection do you see between the Fibonacci sequence and the golden ratio?

Design of the United Nations Building reflects the golden ratio.

3. When a new car is purchased, its value usually decreases with each year of use. Different cars *depreciate* in value at different rates. For example, a new truck purchased for $20,000 might lose 20% of its value each year; a luxury car purchased for $50,000 might only depreciate at a rate of 15% per year. Experiment with different new-car prices and different depreciation rates to find several combinations that will leave you with a car whose value is at least $2,000 after 10 years.

4. In previous courses, you solved systems of linear equations in two variables using several different methods, including the linear-combination method. The general form for such a system in variables x and y with real number coefficients a, b, d, and e and constants c and f is

$$ax + by = c$$
$$dx + ey = f$$

Use what you have learned about solving linear systems of this type to build a spreadsheet algorithm that will accomplish the following:

a. Use input values of the coefficients to test whether the system has no solution, one solution, or infinitely many solutions. Include a statement about what each of the first two cases implies about the graphs of the equations.

b. Give the unique solution when there is one.

Checkpoint

In this unit, you modeled problem situations using spreadsheet software as a mathematical tool.

a For what types of mathematical problems does it seem beneficial to use a spreadsheet?

b In designing a spreadsheet for a particular problem, what general strategies are useful in the planning stage?

c Once a spreadsheet is constructed, what strategies do you use to check its accuracy?

Be prepared to discuss your strategies and thinking with the entire class.

On Your Own

Write, in outline form, a summary of important spreadsheet operations and functions that are useful in mathematical problem solving. Organize your summary so that it can be used as a quick reference in your future work.

Index of Mathematical Topics

A

Abel, Neils, 389
Absolute maximum, 369
Absolute minimum, 369
Absolute value
   of complex number, 399
   in solving equations, 663
Acceleration, *5–6, 17–18*
Acceptance sampling, *329*
Accumulated change, *52–71*
   estimating from a graph,
   53–54
Accuracy in information, 628
Addition of vectors, *87–94, 102*
Additive inverse
   of complex numbers, 392
   in modular arithmetic, 610
   of real numbers, 610
Agnesi, Maria Gaetana, 415
Algebraic properties for real
   numbers, 394
Al-Kwarizmi, 385
AND (logical operator), 580
AND NOT (logical operator),
   580
Angles
   initial side of, 459
   terminal side of, 459
Angular velocity, *123*
ARCLP prime tester, 623
Area, 669
   of equilateral triangle, 668
   in evaluating definite
    integrals, *61–62*
   of similar figures, 668
Arcs, *119–122*
Arithmetic, modular, 607–625
Arrows, *83*
Associative Property
   of addition, 394
   of multiplication, 394
Asymptote, 408
   oblique, 410–411
Authentication, 601
Axis of rotation, 558

B

Bar code, 628–639
Barrow, Isaac, 415
Bernoulli, Jacob, *267*
Bernoulli inequality, *267*
Binary digits (bits), 0 and 1,
   630
Binary operations, 592
Binomial distributions
   approximately normal
    guidelines, *305*
   characteristics of,
    300–315

constructing, using
   simulation, *277–282*
graphs of, *300–307*
mean of
   formula, *307–310*
   as sample size increases,
    302–303, 306
   normal approximation,
    318–322
   sample-size guideline, *280*
   standard deviation of
    formula for, *307–310*
   as sample size increases,
    302–303, 306
   tail of, *278*
Binomial expansions, *245,*
   249, 293
   combinations and, *247*
   Pascal's triangle and, *246*
Binomial expressions, *245*
Binomial population code,
   295
Binomial probabilities
   calculator to compute, *288*
   counting methods to
    compute, *282–286*
   formula, *282–289, 319*
   mean formula, *307–310*
   standard deviation
    formula, *307–310*
   using normal distribution
    to approximate,
    318–322
Binomial situations, *276–297*
   chance variations,
    318–329
   compared to waiting-time
    situations, *281*
Binomial Theorem, *245–249*
Bisection algorithm, 403
Boolean algebra, 597
Boolean operators, 580–581
Box plot charts, *304*

C

Caesar cipher, 603
   security of, 605
   using linear functions
    with, 620
Cardano, 391
Cartesian product sets, 592,
   595
Cause and effect, proving,
   332–351
Chance variation, *318–329*
Check digits, 629–635
   Verhoeff code and,
    642–643
   working of, 630
Ch'in Chiu-shao, 384

Chu Shih-chieh, *246,* 384
Ciphers
   Caeser, 603
   Hill, 604
   substitution, 603–604
Circles, *119–122,* 527
   area of, *151*
   standard form of
    equation of, 528
Circular motion, *119–123*
Circular paraboloid, 525
   cross sections of, 525
Closure Property
   of addition, 394
   of multiplication, 394
Codes, 600–639
   definition of, 639
Coefficients, 362
   method of undetermined,
    366–368
Combinations, 222–231,
   245–249
   $C(n, k)$, 229, 255
   coefficients of binomial
    expansions and, *247*
   counting formulas for, 228
   features on a calculator,
    229
   notation for, *234*
   Pascal's triangle and, *247*
   permutations and,
    228–229
   properties of, 248
Combinatorics, 217, 235
Common logarithms,
   159–163
   converting natural
    logarithms and, 453
   logarithmic scales and,
    168–172
   re-expressing equations,
    164–168
Commutative Property
   of addition, 394
   of multiplication, 394
   of vector addition, *88, 102*
Complement of a set, 586
Complete graph, *253, 261*
Completing the square, 528
Complex conjugate, 503
   properties of, 503
Complex number plane, 399,
   494
Complex number roots
   finding, 498–501
   as a set, 593
Complex numbers, 392, 494,
   593
   absolute value of, 399, 495
   addition of, 392
   additive identity of, 392
   additive inverse of, 392
   algebraic properties of, 394

changing between standard
   form and trigonometric
   form of, 496
conjugates of, 393
connection between
   vectors and, 399
geometric transformations
   and, 496, 505
multiplication of, 393
multiplicative identity of,
   393
multiplicative inverses of,
   393
operations of, and
   geometric
   transformations,
   496–497, 502
polar coordinates of, 496
product of, and its
   conjugate, 393
product of two, 393
quotient of two, 393
relationship between
   complex roots of
   polynomial functions
   and, 392
relationship between
   position vectors and, 495
solving equations
   involving, 401
subtraction of, 392
triangle inequality for, 495
trigonometric form of, 496
Complex roots of real
   numbers, 499–500
Component analysis, *92–93*
Components of a vector, *90*
Composition of functions,
   153–154
   $g(f(x))$, *153*
   noncommutativity of, *154*
Computer Algebra System
   (CAS), *245*
   using, to solve equations,
    490, 511
Conditional probability, *252*
Cone
   cross sections, 516–517
   double, 527, 556, 560,
    568
   surface of revolution, 559
Confidentiality, 601
Conic graph paper
   in drawing ellipses, 537
   in drawing hyperbolas, 538
Conic sections
   identifying, from
    equations, 529, 534, 574
   in terms of intersection of
    a plane and a double
    cone, 527
   writing equation from
    graph, 541, 574

Conics, 527
Conjugate, 393
Constant of proportionality, *206*
Continuous distribution, *328*
Contour diagrams, 516, 542
  interpreting, 517
Contour lines, 514
Contour maps, 514
  interpreting, 520, 541, 572
  making, 519–520, 573
  relationship between relief maps and, 524
Control charts, *327*
Coordinate planes, 522, 539
  equations of, and planes parallel to them, 548
Coordinates and vectors, *94–98, 101–103*
Correlation coefficient
  indicating whether or not a linear model is appropriate, *205–206*
  interpreting, *192*
Cosecant (csc) function, 462
  domain, range, and period of, 462
  graph of, 462
Cosine (cos) function
  circular function, 459
  derivative of, *36*, 477
Cosine(s)
  double angle identity, 473
  finding exact values, 470
  Law of, 92
  right triangle ratios, 89
  sum and difference identities for, 467–471, 476
Cotangent (cot) function, 462
  domain, range, and period of, 462
  graph of, 462
Counterexample, *259*
Counting
  methods of, *216–237*
  Multiplication Principle of, *220, 242*
  organized lists in, *218*
  throughout mathematics, *240–255*
  tree diagrams in, *218*
  when repetitions are allowed, *222–225, 237*
Cross sections, 516, 523–527, 552–558
Cryptography, 600–601
Cryptosystems, 602
  fixed-shift letter-substitution, 603
  hybrid, 616–617
  public-key, 611–619
  ROT13, 602
  symmetric-key, 602–606
Cube roots of unity, 499
Cycle graph, *253*
Cylinder, cross sections of, 525
Cylindrical surfaces, 559–560
  elliptic, 561
  logarithmic, 561
  parabolic, 560

Damped oscillations, 425
Data
  linearizing, *180–207*
  transforming to determine appropriate model, *180–207*
Deceleration, *5–6, 9*
Decibel scale, *169*
Degree, measure, *121*
Degree of polynomial, 362
DeMoivre, Abraham, 498
DeMoivre's Theorem, 498–499
  proof of, using algebraic reasoning, 505
  proof of using Principle of Mathematical Induction, 499
Derivative, *32*
  decreasing, *38*
  dy/dx, *47*
  estimate using difference quotient, *34*
  estimating by slope of graph, *29–31, 39*
  exponential function, *36*
  function f', *33*
  functions, *32–35*
  of $f(x) = \ln x$, 447
  graphs from function graphs, *37–40*
  increasing, *38*
  linear function, *34–35*
  negative, *38*
  positive, *38*
  power function, *36*
  quadratic function, *34–35*
  rules for linear and second degree power functions, *47*
  sums of functions and, *49*
  trigonometric function, *36*
  zero, *38*
Determinant of a 2×2 matrix, 621, 625
Difference identities
  for cosine and sine functions, 467–471, 476
  for tangent functions, 473
  using, 470
Difference quotient, *34*
Differentiation, *46, 69*
Digital signatures, 617
  characteristics of, 619
Dihedral group, 642
Directed line segments, *83*
Direction (heading), *83*
Direction of vector, *95*
Discrete distribution, *328*
Disjoint set, 587
Distributive Property of multiplication
  over addition, 394
  scalar multiplication over vector addition, *88, 102*
  over subtraction, 394
Dot product, *105*
Double cone
  equation of, 560, 568–569
  plane intersections of, 527, 542
Doubling time for investments (Rule of 72), 652

Edge coloring, *253*
e, 445
  defined as infinite series, 454
Electronic information, accuracy in transmission of, 628
Ellipses, 527
  drawing with conic graph paper, 537
  equations for, 529, 531–532, 555
  foci of, 531, 537
  geometry of, 532
  horizontal axis of, 529
  locus-of-points definition of, 531
  major axis of, 529
  minor axis of, 529
  sketching, using standard form of equation of, 529
  standard form of equation of, 529
  symmetry of, 532, 555
  vertical axis of, 529
Ellipsoid, 555, 568
  equation for, 555
Elliptical orbits, *125–126, 133*
Elliptical paraboloid, 568
Elliptical paths, parametric equations for, *125–126, 133*
Elliptic cylindrical surface, 561
Empty set, *230*, 587
Equal sets, 579
Equations, substitution of variable to help solve, 402, 443
Equivalent mod n, 607, 620
Erdös, Paul, *240*
Error function, 402
Errors, correcting and detecting, 628–634
Euclidean algorithm for finding multiplicative inverse in Z_n, 625
Euler, Leonhard, 504
Euler's formula, 504
Euler's Theorem, 614
Even function, 474
Experiments, characteristics of, *333–338*
  clinical trial, *346*
  comparison group, *334*
  control group, *334*
  double-blind, *335*
  evaluator-blind, *335*
  random assignment, *334*
  replication, *334*
  subject-blind, *335*
  subjects, *333*
  treatment, *333*
Exponential equations
  solving, 440–443, 511, 663
  logarithms in, *166–168*
  substitution of variable to help in, 443
Exponential functions, *25*, 437–440, 689
  derivative of, *36*, 454
  domain and range of, 438

doubling time for, 442
e and the derivative of, 454
  estimating instantaneous rates of change of, *25*
  growth rates and, 438, 672
  inverses for, *159–164*
  linearizing data and, *182*
  log transformations and, *186–189*
  recursive formula for, 438
  rewriting in equivalent forms, 438, 451
Exponential regression
  equations and technology method, *187*
Exponents
  properties of, *164*, 439
  relationship between logarithms and, 453

Factored form
  of polynomial, 362
  of quadratic function, 361
Factorial, *226*
Factor Theorem, 386
  proof of, 400
Fermat's Little Theorem, 609
  variation of, to test for primes, 622
Fibonacci sequence, 689
Finite sets, 596
Fisher, R.A., *339*
Fisher's exact test, *349*
Fixed-shift letter-substitution cryptosystem, 603
Four-dimensional analog of a cube, forming, 567
Four-dimensional model, 566
  analog of a sphere in, 566
  distance between points in, 566
  midpoint of a segment in, 566
Fractals, 666–670
Fractal tree, *151–152*
Frequency distributions
  histogram of, *278*
  for the number of successes compared to proportion of successes, *281*
Frequency table, *277*
  in estimating probability, *278*
Function formulas, recursive formulas and, 667–670
Function graphs, translation of, 408
Function(s)
  absolute maximum and minimum, 369
  composition, *145, 153–154*, 425
  decreasing, *154*
  definition of, in terms of ordered pairs and sets, 596
  derivative of, *37–40*
  difference quotient for estimating the derivative, *34*

end behavior, 368, 371
exponential, *25, 36, 160*
finding equation of
inverse of a, *148*
increasing, *154*
inverse cosine, *155*
inverse sine, *155*
inverses, *142–155*
linear, *23–24, 34*
linearizing, *180–207*
definition of, *182*
local maxima and minima, 369
logarithmic, *158, 163*
multiplicity of zeroes, 370
natural log, 444–447
one-to-one, *144*
piecewise-defined, *155*
power, *36*
quadratic, *24–25*
square root, *183*
trigonometric, cosecant, cotangent, secant, 462
Fundamental theorem of algebra, 392, 395, 494
$f'(x)$, *32*
estimating by using difference quotient for, *33–37, 47*
estimating from graph of $f(x)$, *37–40*
using to find characteristics of the graph of $f(x)$, *31, 43–44*
$f(x)$
patterns in and rate of change of, *27*
properties of $f'(x)$ and the graph of, *43*
rule in approximating slope of graph, *43*

G

Galois, Evariste, 389
General Multiplication Rule to find probability of any two events, *243*
Geometric mean, 686
Geometric series, *266, 274*
Geometric transformations
complex numbers and, 504–505
inverses of, *154*
relationship between multiplication of complex numbers and, in plane, 496
Golden Ratio, 689
Graphs
binomial distributions, *301–315*
complete, *253, 261*
cycle, *253*
derivative from function graphs, *37–40*
edge coloring, *253*
estimating rate of change, *11*
function inverse, *147–150*
rate of change, using to estimate net change, *56–59*

velocity graph used to sketch distance, *31*

H

Half-angle identities, 477
Heading, *82*
Hill cipher, 604
general version of, 621
security of, 605
Histograms
estimating mean and standard deviation, *278*
of frequency distribution, *278*
Horizontal asymptote, 408
Horizontal line test, *152*
Horizontal planes, equations of, and planes parallel to them, 548
Horner, William George
Horner's Method, 384
Hybrid cryptography, protocol for, 617
Hyperbolas, 527
asymptotes of, 533
branch of, 533
drawing with conic graph paper, 538
equations of asymptotes of, 533, 543
foci of, 533, 538
locus-of-points definition of, 532
parametric equations for, 533, 543
standard forms of equation of, 533
symmetry lines of, 533
vertices of, 533
Hyperbolic paraboloid, 569
Hyperboloid
of one sheet, 568
of two sheets, 569
Hypercube, 567
Hypergeometric formula, *296*

I

Identities, half-angle, 477
Imaginary number i, 391, 494
Imaginary numbers, find roots of, 499
Imaginary part of $a + bi$, 399
Independent events, multiplication rule for, *242*
Induction, Principle of Mathematical, *261*
Infinite series, 454
Infinite sets, 596
having same number of elements, 596
Infinity, approaching, 368
Informatics
access, 578–588
accuracy, 628–635
security, 600–606, 611–619
Information digits, 629
Initial side of angle, 459

Inner product, *105*
Instantaneous rates of change, *2–19*
using $f(x)$ rule to estimate, *27*
Instantaneous velocity, 421
Integers
as a set, 593
mod n, 607–611
sum of first n odd positive, *263*
Integrals, *60–64*
area in estimating, *61–62*
calculator or computer routine in evaluating, *62*
of $f(x)$ from a to b, *61*
rectangles in estimating, *60–63*
Integration, *69*
Intercepts
of a three-dimensional surface, 556, 574
Interest rates, nominal and effective, 446
Intersection (∩), of sets, 582
Inverse functions, *142–155*
cosine, *155, 482*
evaluating from tables, *145*
function compositions, *145, 153*
graphs, *147–150*
parametric equations and, *155*
property of f and f^{-1}, *145*
relationship between range of f and domain of f^{-1}, *146*
sine, *155, 482*
Inverse variation, 430
Irrational numbers, 390, 593

K

Kepler's Third Law, *200*
Key for encryption, 603
Kline, Morris, *131*
Koblitz, Neal, *138*

L

Law of Cosines, using the, *92*, 469
Law of Sines, using the, *105*
Least squares, method of, *195*
Least squares regression line and residuals, *186*
Leibniz, Gottfried, *46–47*
Li Chih, 384
Limiting values, 667–669
Linear-combination method, solving systems using, 567
Linear equations
interpreting and solving of systems of three, in three variables, 567
linear combinations for solving systems of, 367
matrix method for solving systems of, 367, 567

solving, with spreadsheets, 661–662
solving systems of, with spreadsheets, 690
in three-space, 552–553
Linear factor of polynomial, 362
Linear functions, *23–24*
with arithmetic mod 26, 620
derivative of, *34*
inverse of, *147–149*
inverse of, piecewise, *155*
finding slope of, and estimating rate of change of, at a point, *29–31*
Linear inequalities, solving with spreadsheets, 661–662
Linearizing data, *180–207*
exponential functions, *182–189*
power models, *182, 199*
square root functions, *182, 184*
translated power functions, *207*
Linear motion
modeling, *80–105*
parametric models simulating, *109–115*
Linear scale, *159*
Linear form
logarithmic equations, solving, 448
trigonometric equations, solving, 481
Local linearity, *29–32*
Local maximum, 369
Local minimum, 369
Locus-of-points perspective
for ellipse, 530
for hyperbola, 532–533
for parabola, 530
Logarithmic cylindrical surface, 561
Logarithmic equations
extraneous solutions to, 450
solving, *162–163*, 448–450
Logarithmic expressions, evaluating, *162*
Logarithmic functions, *158–163*
comparison of graphs $g(x) = \log x$ and $f(x) = \ln x$, 447
derivative of, *175*, 445
domain and range of, 442
graphs of, *161–162, 175*, 442
graphs for various bases, *175*, 444–445
Logarithmic scale, *158, 168–172*
Logarithms
change of base formula, 444
common, *159–164*
Power Property, *166*
Product Property, *165*
proof using mathematical induction, *266*

Index of Contexts

Photo Credits